VERY

BRITISH

WEATHER

Ebury Press, an imprint of Ebury Publishing
20 Vauxhall Bridge Road
London SW1V 2SA

Ebury Press is part of the Penguin Random House
group of companies whose addresses can be
found at global.penguinrandomhouse.com

Penguin
Random House
UK

First published by Ebury Press in 2020

www.penguin.co.uk

A CIP catalogue record for this book is
available from the British Library

ISBN 9781529107616

Words by Paul Murphy
Page design and typesetting by Emily Voller

Illustrations by Emily Voller and Hannah Fleetwood

Printed and bound in Italy by Printer Trento S.R.L.

MIX
Paper from
responsible sources
FSC® C018179

Penguin Random House is committed to a
sustainable future for our business, our readers
and our planet. This book is made from Forest
Stewardship Council® certified paper.

Image Credits

As below, all other images © the Met Office 2020.

Aidan McGivern, page 35 photo 2, page 53 photo 6.
Andrew Tedd, page 66.
Claude Monet, page 141.
Dan Harris, page 236.
Darren Hardy, page 95.
Graeme Whipps, page 50 photo 2.
Graham Fraser, page 193 photo 1 and 7.
Hazel Bremner, page 135.
Jamie Urquhart, page 193 photo 5.
John Hallam, page 35 photo 4.
Kathleen MacLeod, page 193, photo 3.
Katie McGivern, page 19 'Autumn Leaves', page 35 photos 1 and 3.
Krakatoa Committee of the Royal Society, G.J. Symons (editor), page 222.
Matt Clark, page 35 photos 5 and 6, page 50 photo 1, page 53 photos
    2, 3, 4 and 5, page 193–4 photos 2, 4, 8 and 9.
NASA, pages 113, 215, 231, 237.
Nick Krol, page 53 photo 1.
Phillip Normanton, page 50, photo 4.
P.J.B. Nye, page 50 photo 3.
TORRO Hailstone Intensity Scale, page 92.
Vlad Ispas, page 193 photo 6.

VERY

BRITISH

WEATHER

Over 365 Hidden Wonders from the World's Greatest Forecasters

 **Met Office**

EBURY
PRESS

CONTENTS

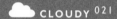

❄ FREEZING ¹⁸¹

❄ SNOWY ¹⁹⁹

❄ FREAKY ²¹⁹

Foreword

Weather has exercised a fascination for humans through recorded history, and nowhere more than in our islands with their changeable conditions. For almost 170 years, Met Office scientists have been driven by their curiosity to better understand the mysteries of weather and to predict it to the benefit of their fellow citizens. In 1943 during the Second World War *Weather*, a predecessor to this book, was published. A year later, the Met Office made a weather forecast that historians argue changed the course of the war.

Much has changed since then. It is startling how ubiquitous and accurate weather forecasts have become; on our phones, the web and TV. *Weather's* closing paragraphs suggest that it may never be possible to forecast next week's weather. However, the work of meteorologists and scientists around the world – combined with ever more powerful supercomputers – mean the dreams of 1943 are now reality.

Another change is that the world has warmed by around a degree Celsius. As a result, we are seeing more weather extremes. In 2019, the UK recorded its highest temperature on record. In 2020, the wettest February on record in the UK brought devastating floods. It is more crucial than ever that we provide accurate forecasts and advise on the impacts of severe weather.

In February 2020, the Met Office received funding for a new supercomputer which will be operational from 2022. Together with our world-leading scientists and meteorologists, this will ultimately deliver earlier and more accurate warning of severe weather. It will also provide the information needed to build a more resilient world and help support the transition to a low carbon economy across the UK.

But you don't need a powerful supercomputer to appreciate watching the weather from your window and wonder why it is so weird and wonderful. We all experience the weather. This book will help you to enjoy and under-stand it. We are proud to share all that we have learned over the years and this book will let you indulge your curiosity about the weather with us.

Professor Penny Endersby
Chief Executive, Met Office

The Met Office: From 1854 to Now

1854 Met Office established by Vice Admiral Robert FitzRoy within the Board of Trade.

1859 Royal Charter storm: approximately 459 lives lost off Anglesey and leads to introduction of a storm warning service.

1861 Newspapers take first public weather forecasts.

1861 Warning service for shipping delivered by telegraph to harbour towns.

1916 First military operational forecast.

1919 WWI has just ended and the Met Office becomes part of the Air Ministry.

1922 First daily weather forecast is broadcast on radio.

1924 First Shipping Forecast is broadcast on radio.

1936 First televised weather maps.

1944 The Met Office advises on Operation Overlord and provides key forecasts for D-Day.

1954 First in-vision weather forecast is presented by George Cowling.

1960 TIROS – the world's first meteorological satellite – is launched.

1962 The Mobile Meteorological Unit is established to provide support for military exercises across the world.

1965 First operational forecast by a computer nicknamed 'Comet'.

1986 NAME – the Met Office atmospheric dispersion model – is developed in response to the Chernobyl nuclear disaster.

1987 The Great Storm occurs and results in the National Severe Weather Warning Service.

1988 The Meteorological Office becomes the Met Office.

1990 The Hadley Centre for Climate Prediction and Research is founded.

1995 Met Office launches its website.

2003 Met Office Headquarters move from Bracknell to Exeter.

2009 Flood Forecasting Centre is founded, a joint operation between the Met Office and the Environment Agency.

2009 Met Office joins Twitter. This is followed by Facebook in 2010, Instagram in 2013, Snapchat in 2017 and TikTok in 2019.

2010 The London Volcanic Ash Advisory Centre, based at the Met Office, monitors and forecasts ash dispersion of erupting Icelandic volcano Eyjafjallajökull.

2011 The Met Office moved from the Ministry of Defence to the Department for Business, Innovation and Skills.

2014 The Met Office produces its first operational space weather forecasts.

2015 Storm Abigail hits north-west Scotland and is the first storm to be named by the Met Office.

2016 Met Office creates an App for iPhone and Android.

2020 Met Office secures £1.2 billion of funding for a state-of-the-art supercomputer that will improve severe weather and climate forecasting. Computing capacity is expected to increase six-fold when it launches in 2022.

A VERY BRITISH OBSESSION

Sunshine is delicious, rain is refreshing, wind braces us up, snow is exhilarating; there is really no such thing as bad weather, only different kinds of good weather.

JOHN RUSKIN

What do we think of when we think about the British weather? Unseasonal heatwaves in April; gale-force winds battering us in November; bright, clear days in January when the sun is shining but it is bitterly cold; warm, sultry days in October as we enjoy an Indian summer; overcast, drizzly days in…well, really at any time of the year. The answer is, of course, all of these and more.

We experience a wide range of seasonal weather in the UK, and it is no exaggeration to say that the British are obsessed with it. It has inspired some of our greatest artists, poets, musicians and writers, and informed our way of life on this small group of islands on the western edge of Europe in the north-east Atlantic Ocean, from the clothes we wear to the buildings we work and live in. Our language is rich with words to describe the weather, and the regional variance is as diverse as the weather conditions we experience. It is also one of our favourite topics of conversation. But why is this the case? Why do we in the UK care so much about the weather?

One potential reason is our geographical position. The British Isles sit in the midlatitudes, where warm air from the subtropics often clashes with cold air from the North Pole. Add to that an ocean to our west and a continent to our east, and you could argue that the UK doesn't really experience its own weather at all, but instead borrows it from other climes: south-westerlies in the autumn bring Azorean levels of humidity; easterlies in the winter carry a Siberian chill; and southerlies in the summer provide a taste of Spanish heat.

Our expertise in the weather also owes much to our island status and resulting maritime history. The Met Office was established in 1854 by Vice Admiral Robert FitzRoy (1805–65) as a department of the Board of Trade with the aim of providing information about the weather to mariners. (FitzRoy also found fame as captain of HMS *Beagle* on Darwin's extraordinary expedition.) Since then, the Met Office has become one of the foremost authorities on all things weather related and a world leader in the science of weather forecasting. As a result of climate change our global weather is in flux as never before, and as a consequence the role of weather forecasting could not be more important.

Thankfully, our understanding of the weather and weather-prediction technologies are improving all the time. Four-day forecasts are as accurate today as one-day forecasts were in 1980. Met Office advice is now increasingly focused on the impacts of its weather forecasts and climate research, helping people to make the right decisions to stay safe and thrive – from helping shops plan what items to stock ahead of a bank holiday weekend, to preventing disruption from fog at the UK's airports and warning of devastating flooding to communities. And future investment in supercomputing during the next ten years will allow us to improve our forecasts further and tailor our advice like never before.

As you can see, the weather is central to life in Britain. But very few of us know why we experience the weather we do, or why it is so important that we are able to predict it accurately. Join us, then, on a journey through a year in the life of the British weather. Starting in spring and ending in winter, we will visit the full range of weather conditions, from the scorching heights of summer to the snowy days in winter when our country comes to a standstill. Of course, we are just as likely to get drenched by a downpour in summer as enjoy a beautifully sunny day in the middle of winter, so this journey will not be without its twists and turns. We will find out when the British weather has been at its best and worst, and a few myths will be busted along the way. And by the end, we will have a much better understanding of this very British obsession.

How to Use this Book

Depending on what the weather is doing each day, this book is designed so that you can look up whichever particular section is relevant – be it rain or shine, frost or fog. There are more than 365 facts to enjoy through the 229 sections in the book. Dip in and out as the days go by, or enjoy reading up on meteorological marvels and top trivia no matter what is happening outside your window.

In each section you'll discover the how, what, why and wow of all things British weather from experts at the Met Office. There are some activities you can take part in yourself, historical gems from the archive, fascinating science and much more. As well as inspiring you to take a deeper look at the sky above us, you'll learn how to *really* talk about the weather in more ways than you had ever imagined!

changeable [cheyn-juh-buhl]

adjective weather that is liable to change; variable; fickle

origins: 1175–1225; Middle English *cha(u)ngen*; Anglo-French, Old French *changer*; Late Latin *cambiāre*, Latin *cambīre* 'to exchange'

CHANGEABLE

It is often said that we can experience four seasons in one day in the British Isles, and while this might be exaggerating things a bit, it is true that our weather can be highly changeable. This is in large part down to the geographic position of the UK. We are an island nation, north of the equator, with an ocean to the west of us and a large landmass to the east. Britain is also located in an area where six main air masses meet, and as a result of these clashing air masses we experience a wide range of frontal systems and their associated weather conditions. Think of it like people running to the same spot from different locations, and having to wrestle it out! An air mass is a large body of air that has similar temperature and moisture properties throughout. Most of the frontal systems that affect the UK are formed out over the Atlantic Ocean and move in a west to east direction. Plus, there is the jet stream (more on that later). This unique set of geographic and atmospheric circumstances makes the British weather as endlessly changeable as it is. It can be difficult to forecast at times, which is why the expertise at the Met Office is so important. There's never a dull moment forecasting the great British weather!

Weather Fronts

The air around the globe has different properties. Air from the north is generally colder, and there is usually a sharp boundary where this cold air meets warmer air from the south, rather than a gradual increase in temperature. Also, air is often dry if it has come from a large continent or holds a lot of moisture if it has come across an ocean.

When two air masses with contrasting properties meet, these differences produce a reaction (often a band of rain) in a zone known as a front. The strength of a front is often determined by how significant the contrast is between the two air masses. If very cold air comes into contact with warm tropical air, the front can be 'strong' or 'intense'. However, little difference in temperature between the two air masses may create a 'weak' front.

ARCTIC MARITIME AIR MASS
From: Arctic
Wet, cold air brings snow
showers in winter

**POLAR MARITIME
AIR MASS**
From: Greenland/
Arctic Sea
Wet, cold air brings
cold showery
weather

**POLAR CONTINENTAL
AIR MASS**
From: Central Europe
Very cold air in winter
brings snow showers
to the east coast

**RETURNING POLAR
MARITIME**
From: Greenland/Arctic
Sea via North Atlantic
Moist, mild and unstable
air bringing cloud
and rain showers

TROPICAL MARITIME AIR MASS
From: Atlantic
Warm, moist air brings cloud,
rain and mild weather

**TROPICAL CONTINENTAL
AIR MASS**
From: North Africa
Hot, dry air brings hot
weather in summer

COLD FRONTS

A cold front is symbolised on a weather map by a line with triangles. The triangles can be thought of as icicles, and they are usually coloured blue. The presence of a cold front means that cold air is advancing and pushing underneath warmer air. This is because the cold air is heavier, or denser, than the warm air. The tips of the triangles indicate the direction of movement of the cold air.

WARM FRONTS

A warm front is symbolised on a weather map by a line with semicircles. The semicircles can be thought of as half suns, and they are most often coloured red. The presence of a warm front means that warm air is advancing and rising over cold air. This is because warm air is lighter, or less dense, than cold air. The edges of the semicircles indicate the direction of movement of the warm air.

OCCLUDED FRONTS

An occluded front is symbolised on a weather map by a line with both semicircles and triangles. They are often coloured purple. An occlusion can be thought of as having the characteristics of both warm and cold fronts.

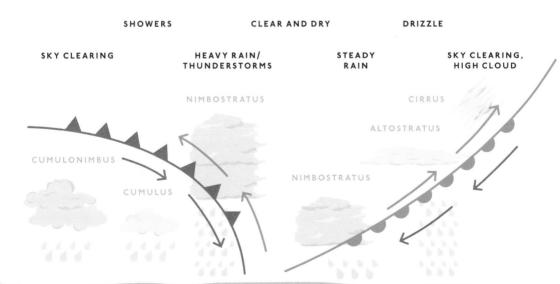

From Hot to Cold in Twenty-four Hours

Because of our varied weather conditions, the contrast in temperature across a 24-hour period can be quite pronounced. This daily difference between the highest and lowest temperatures is referred to as diurnal range, and in the UK this can be especially significant in the spring when the skies are clear. The land remains cool following winter and temperatures can quickly fall after dark. It is also the time of year when the power of the sun is beginning to be felt more strongly again in the northern hemisphere. As a result, strong sunshine can soon turn frosty mornings into warm afternoons.

Although diurnal range is not something that is measured specifically, in the way rainfall is, the largest temperature range in a single day found in the Met Office's database was recorded on 14 January 1979, when Lagganlia (near Aviemore) went from -23.5°C to 6.6°C over a 24-hour period, a change of 30.1°C. The average diurnal temperature range for the UK (the difference between the average maximum and average minimum temperatures) is 7.2°C, and the highest average diurnal temperature range is in May (8.8°C) and the lowest is in December and January (5.5°C).

OH, THOSE HOT WINTER NIGHTS... ?

Surprisingly, it is also very common for nights to be warmer than daytime temperature, particularly in winter when a change in air mass can have a more important effect than daytime heating. One significant recent example would be the record-breaking maximum temperature for December, with 18.7°C observed at Achfary in Sutherland at around 2am on 29 December 2019.

Highs and Lows from Shetland to Surrey

In terms of place, Surrey has the highest average diurnal temperature range in the UK (8.5°C) and Shetland the lowest (4.3°C). This may be down to the fact that Shetland is surrounded by sea and experiences stronger winds compared to Surrey, both of which can help moderate its climate. Surrey is inland and, because of other factors such as topography and soil type, can get quite cool under clear skies overnight. Since Surrey is in the south-east of the UK it can also get quite hot on a summer's day, hence the wider range. In Shetland, there is a significant influence on the weather from the surrounding seas, which respond more slowly to temperature change. So, overnight temperatures won't get down as far as more landlocked areas, and during the day, because of its far-north latitude, it is less often hot in Shetland.

How Extreme Do We Get in the UK?

As well as the potential for great variance in weather and temperature in a single day, the changeable nature of the British weather is such that we also experience a wide range of variance throughout the year. This contributes to temperatures that reach into the high thirties during the summer and the minus tens and twenties during winter. The highest temperature on record in the UK is 38.7°C, recorded in Cambridge in 2019, and the lowest is -27.2°C, recorded in Braemar in Scotland on 11 February 1895 and 10 January 1982, as well as at Altnaharra, also in Scotland, on 30 December 1995. This is the equivalent of 65.9°C between the two extremes.

This vast range in temperatures through the year can be problematic, especially for infrastructure. It is difficult to build roads and railways that can perform consistently in such different temperatures. Tarmac can melt in the heat and break up under freeze-thaw action, and railway lines can buckle in very high temperatures. It also means that the wardrobes of the Great British public vary widely, from big warm coats to T-shirts and shorts – something people in other parts of the world don't have to worry about so much!

COUNTRY	TEMP (°C)	DATE	LOCATION
England	38.7	25 July 2019	Cambridge Botanic Garden
Wales	35.2	2 August 1990	Hawarden Bridge (Flintshire)
Scotland	32.9	9 August 2003	Greycrook (Scottish Borders)
Northern Ireland	30.8	30 June 1976	Knockarevan (County Fermanagh)
Northern Ireland	30.8	12 July 1983	Shaw's Bridge, Belfast (County Antrim)

THE CHILLY CHAMPIONSHIPS

COUNTRY	TEMP (°C)	DATE	LOCATION
Scotland	-27.2	10 January 1982	Braemar (Aberdeenshire)
Scotland	-27.2	11 February 1895	Braemar (Aberdeenshire)
Scotland	-27.2	30 December 1995	Altnaharra (Sutherland)
England	-26.1	10 January 1982	Newport (Shropshire)
Wales	-23.3	21 January 1940	Rhayader (Powys)
Northern Ireland	-18.7	24 December 2010	Castlederg (County Tyrone)

What is the Jet Stream?

The jet stream is a core of strong winds around 8 to 11 kilometres above the Earth's surface, blowing from west to east.

HOW DOES THE JET STREAM WORK?

The jet stream flows high overhead and causes changes in the wind and pressure at that level. This affects conditions nearer the surface, such as areas of high and low pressure, and therefore helps shape the weather we see. Sometimes, like in a fast-moving river, the jet stream's movement is very straight and smooth. However, its movement can buckle and loop, like a river's meander. This will slow things down, making areas of low pressure move less predictably.

The jet stream can also change the strength of an area of low pressure. It acts a bit like a vacuum cleaner, sucking air out of the top and causing it to become deeper, lowering the pressure system. The lower the pressure within a system, generally the stronger the wind, and more stormy the result.

On the other hand, a slower, more buckled jet stream can cause areas of higher pressure to take charge, which typically brings less stormy weather, light winds and dry skies.

The Four Seasons

Life in Britain is greatly influenced by the four seasons and by the weather that each brings. Over the years, we have cleaned out our houses in spring, gone on holiday in the summer, harvested our crops in autumn and hunkered down for the winter.

The seasons occur as a result of the 23.5 degrees of tilt of the Earth's rotational axis in relation to its orbit around the sun (our calendar year), meaning certain areas of the globe are tilted towards it while other areas are tilted away from it. This creates a difference in the amount of sunlight (or solar radiation) that reaches different parts of the Earth, which in turn creates the global cycle of fluctuations that we know as the seasons.

There are different ways to define when the seasons start and finish.

Meteorological seasons consist of splitting the year into four periods made up of three months each. These seasons coincide with our Gregorian calendar, making it easier for meteorological observing and forecasting to compare seasonal and monthly statistics. By this method, spring runs from March to May, summer from June to August, autumn from September to November and winter from December to February.

The seasons can also be defined astronomically, using key moments in the Earth's year-long, elliptical journey around the sun.

☀ **APPROX NO. OF DAYLIGHT HOURS IN EDINBURGH AND LONDON**

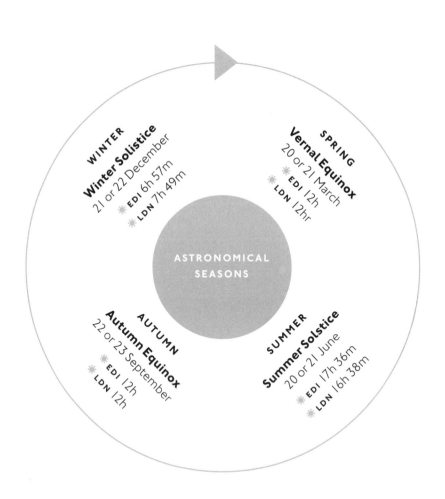

WINTER
Winter Solstice
21 or 22 December
☀ **EDI** 6h 57m
☀ **LDN** 7h 49m

SPRING
Vernal Equinox
20 or 21 March
☀ **EDI** 12h
☀ **LDN** 12hr

ASTRONOMICAL SEASONS

AUTUMN
Autumn Equinox
22 or 23 September
☀ **EDI** 12h
☀ **LDN** 12h

SUMMER
Summer Solstice
20 or 21 June
☀ **EDI** 17h 36m
☀ **LDN** 16h 38m

The Art of Forecasting – and Why It's So Challenging in the UK

The fickle nature of the British weather can at times make it difficult to forecast specifically for individual locations. Although we might be able to accurately forecast weather fronts, there can sometimes be uncertainty about the timing of where and when they will be felt; for example, exactly when rain might fall in a particular place.

What is the Best Way to Stay Ahead of the Weather?

Because the weather in the UK can change from day to day, from hour to hour and even on occasion from minute to minute, it is advisable to check the forecast regularly. This is especially true when low-pressure systems are moving across the UK, as the forecast could develop rapidly as more models are run by the Met's supercomputers. In the same vein, it is always worth watching a Met Office-presented weather forecast, as the presenter will be a meteorologist who will be able to explain any uncertainty and will interpret the data that has been gathered. Nothing beats a skilled human delivering the subtleties of all the data.

The Joys of Change

Our varied and changeable weather has proven to be a fertile source of inspiration for our rich and diverse culture. From the seasons fuelling changing fashions as we move from garments designed to keep us warm in winter to those that keep us cool in summer, to the myriad works of art, music and literature that reference and celebrate every weather condition you can think of, the weather is always with us, and is always influencing how we live. Whether you are a sun worshipper or someone who enjoys skiing down a snowy slope, whether you like to cosy up beside the fire on a stormy autumn evening or find joy in the green shoots of spring, there is no doubt that the very British weather has something for everyone. And if the rain starts to pour on your summer barbecue party? Just put on a waterproof layer and persevere.

Data Is King

Observations are measurements of meteorological quantities, such as temperature, rainfall, pressure, wind and humidity, and weather forecasting has always relied on these figures, particularly in the UK where the weather is so changeable. The word 'observations' derives from the early days of meteorology when all recordings of the weather were made by trained observers. Today, taking observations is a mainly automated process using twenty-first-century technology.

Before we can make any forecast about our varied weather, we need to know what is happening now. We make observations to tell us where there is rain, snow, fog or frost and how severe the weather is. To then make a forecast of what the weather is going to be like, we need to make accurate measurements, including pressure, temperature and the strength of the wind.

It is particularly important that there are observations throughout the depth of the atmosphere that define its three-dimensional structure (*see* 'Showery'). For these reasons, the Met Office invests a great deal of money in observing systems capable of measuring the atmosphere in all its complexity. Weather systems in one part of the world can quickly have an impact on other more distant areas, so an accurate forecast over the UK requires observations not only throughout the UK, but over many other parts of the globe. To get all the observations the Met Office collaborates closely with other countries in the exchange of data.

This data tells us how today's weather differs from the long-term average in the UK climate and how our climate has changed over timescales of decades or centuries. This has become increasingly important as a result of the threats posed by climate change. Accurate observations are required as supporting evidence for the decisions made by government and industry in response to this growing challenge.

Connected World

Because conditions can change so quickly, it is important to be aware of the weather in other parts of the world that might influence the weather here in the UK. Although our weather fronts tend to come in from the west over the Atlantic, sometimes they can switch. If the weather is very cold in continental Europe, for example, and there is a change in the direction of the prevailing weather conditions from west to east, we would expect to experience those cold conditions in the UK as well. The same is true if a front comes in from the south, bringing with it warmer weather from as far away as Africa.

The causes of our fickle climate can also be predicted by looking further afield. Because the UK's weather often comes from the west, looking at what is going on over Canada and the US can inform what the forecast is likely to be for the UK. Interactions with other large-scale force factors such as the jet stream are also important. Depending on where the jet stream is, especially during the autumn and winter months, it can significantly influence how unsettled the weather can be in Britain. If the jet stream is directly over the UK, it can force low-pressure storm systems straight into the British Isles, bringing wet and windy weather. If the jet stream is to the north, we would expect milder and more settled weather. If it is to the south, we would expect cold, potentially snowy conditions.

Types of observations

1 SATELLITE: imaging the whole globe from space

2 LAND SURFACE: measuring the weather that we experience on the ground using instruments

3 MARINE: making measurements near the surface of the ocean and through its depths

4 UPPER AIR: visualising the 3D structure of the atmosphere above the Earth's surface

5 RADAR: providing the fine detail of rainfall on scales down to one kilometre

6 THUNDERSTORM LOCATION: pinpointing thunderstorms and their associated severe weather

The Weather Observations Website received **800 million** citizen observations in its first 5 years

We hold the longest-running temperature series in the world

We record over **30 million** observations per year

Our engineers maintain **10,000** system components from simple rain guages to the latest radar technology

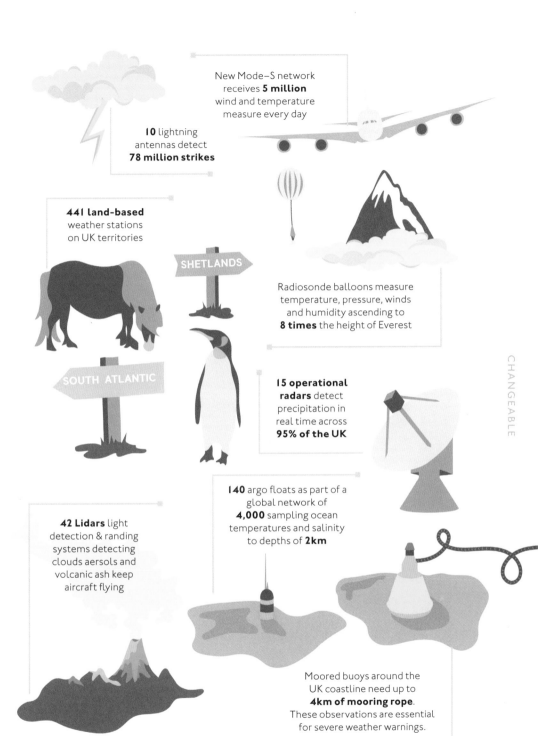

New Mode–S network receives **5 million** wind and temperature measure every day

10 lightning antennas detect **78 million strikes**

441 land-based weather stations on UK territories

SHETLANDS

SOUTH ATLANTIC

Radiosonde balloons measure temperature, pressure, winds and humidity ascending to **8 times** the height of Everest

15 operational radars detect precipitation in real time across **95% of the UK**

42 Lidars light detection & randing systems detecting clouds aersols and volcanic ash keep aircraft flying

140 argo floats as part of a global network of **4,000** sampling ocean temperatures and salinity to depths of **2km**

Moored buoys around the UK coastline need up to **4km of mooring rope**. These observations are essential for severe weather warnings.

CHANGEABLE

Super-duper Computers

All of these observations are not worth much unless they can be interpreted and analysed in such a way as to increase our understanding of how the weather is formed and therefore how we can predict it. Once the huge variety of measurements are made, they provide the initial conditions for the complex computer models that run several times a day on supercomputers that are so powerful they seem like something out of science fiction.

The Met Office has three Cray XC40 supercomputing systems, capable of more than 14,000 trillion arithmetic operations per second – a mind-boggling number equivalent to more than 2 million calculations per second for every man, woman and child on the planet. The computers contain two petabytes (or 2,000 terabytes) of memory, enough to hold 200 trillion numbers, a total of 460,000 computer cores and 24 petabytes of storage for saving data, enough to store more than 100 years' worth of HD movies. This power allows the Met Office to take in 215 billion weather observations from all over the world every day, which it then uses as a starting point for the atmospheric models it runs, each of which contains more than a million lines of code.

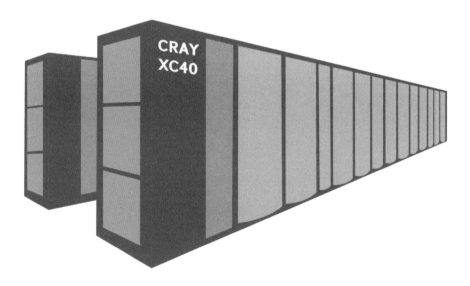

But as with your home computer or smartphone, things change quickly and what is cutting-edge today soon becomes out of date. As a result, the UK government has given the go-ahead for us to replace our existing supercomputing infrastructure in 2022, at the end of its useful life, thereby allowing the Met Office to remain at the forefront of weather forecasting and climate research for the next ten years and beyond.

Lewis Fry Richardson's Forecast Factory

Remarkably, one pioneering mathematician and scientist predicted how computer weather forecasts would work more than twenty years before the first computer was even invented. In 1922, Lewis Fry Richardson (1881–1953) described a large, circular, theatre-like hall, with a map of the world painted on its walls. He estimated that 64,000 'computers' (not computers as we know today, but mathematicians tasked with 'computing') would then be needed to calculate weather forecasts in real time, each responsible for one small part of the globe. Of course, computers were mechanised when they arrived, but Richardson's general concept was proven to be correct.

The First Supercomputer

Although weather-prediction computers were used for research and testing during the 1950s and early 1960s, the first operational computer forecast was produced on 2 November 1965. This was the English Electric KDF9, also known as 'Comet'. Costing £500,000, Comet could perform 60,000 arithmetic operations per second. Forecasts were produced twice a day for the North Atlantic and Western Europe at a resolution of 300 kilometres. We've come a long way since.

Forecasting Firts

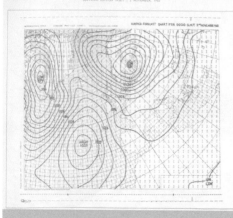

THE WEATHER.
METEOROLOGICAL REPORTS.

Wednesday, July 31, 8 to 9 a.m.	B.	E.	M.	D.	F.	C.	I.	S.
Nairn	29·54	57	56	W.S.W.	6	9	o.	3
Aberdeen	29·69	59	54	S.S.W.	5	1	b	3
Leith	29·70	61	55	W.	3	5	c.	2
Berwick	29·69	59	55	W.S.W.	4	4	c.	2
Ardrossan	29·73	57	55	W.	5	4	c.	5
Portrush	29·73	57	54	S.W.	2	2	b.	2
Shields	29·80	59	54	W.S.W.	4	5	o.	3
Galway	29·83	65	62	W.	5	4	c.	4
Scarborough	29·86	59	56	W.	3	6	c.	2
Liverpool	29·91	61	56	S.W.	2	6	c.	2
Valentia	29·97	62	60	S.W.	2	5	o.	3
Queenstown	29·88	61	59	W.	3	5	c.	2
Yarmouth	30·05	61	59	W.	5	2	c.	3
London	30·02	62	56	S.W.	3	2	b.	—
Dover	30·01	70	61	S.W.	3	7	o.	2
Portsmouth	30·01	61	59	W.	3	6	o.	2
Portland	30·03	63	59	S.W.	3	2	c.	3
Plymouth	30·00	62	59	W.	5	1	b.	4
Penzance	30·04	61	60	S.W.	2	6	c.	3
Copenhagen	29·94	64	—	W.S.W.	2	6	c.	3
Helder	29·99	63	—	W.S.W.	5	5	c.	3
Brest	30·09	60	—	S.W.	2	6	c.	5
Bayonne	30·13	63	—	—	—	9	m.	5
Lisbon	30·18	70	—	N.N.W.	4	3	b.	2

General weather probable during next two days in the—
North—Moderate westerly wind ; fine.
West—Moderate south-westerly ; fine.
South—Fresh westerly ; fine.

Explanation.
B. Barometer, corrected and reduced to 32° at mean sea level ; each 10 feet of vertical rise causing about one-hundredth of an inch diminution, and each 10° above 32° causing nearly three-hundredths increase. E. Exposed thermometer in shade. M. Moistened bulb

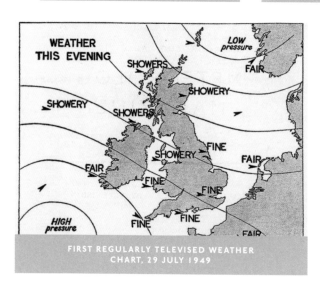

WEATHER THIS EVENING

LOW pressure

SHOWERS

FAIR

SHOWERY

SHOWERY

SHOWERS

SHOWERY

FINE

FAIR

FAIR

FINE

FINE

FINE

FINE

HIGH pressure

FAIR

What is the Difference Between Weather and Climate?

The main difference between weather and climate is that they each refer to different timescales. Weather describes the conditions of the atmosphere, including temperature, rainfall, cloudiness, sunshine or wind speed, over hourly or daily measurements. Climate, on the other hand, is the average of these conditions over longer time periods, ranging from years to decades and even centuries. In the words of writer Robert A. Heinlein, 'Climate is what you expect; weather is what you get'. Perhaps even more simply put: 'Weather is how you choose your outfit; climate is how you choose your wardrobe'.

Is There a Calm Before the Storm?

The common phrase 'the calm before the storm' could be a reference to a weather phenomenon known as ridging. Ahead of low-pressure systems that bring wet and windy conditions, there can sometimes be a brief 'ridge' of higher pressure that brings more settled weather and possibly sunshine, too. Under higher pressure, winds are lighter, and the skies tend to be clear and sunny, giving the impression of calm conditions. However, the settled weather doesn't last long, as ridges of high pressure are fleeting, and areas of low pressure follow closely behind them.

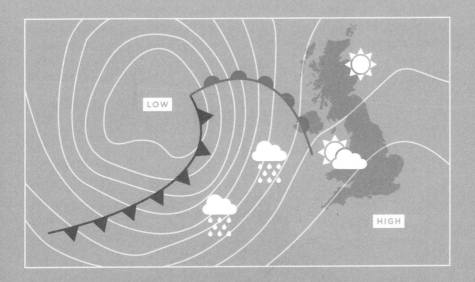

Mountain Safety

The inconvenience of getting caught without a raincoat or an umbrella in an April shower pales into insignificance when compared with being caught out by changing weather in the high mountains of Wales and Scotland. The Highlands, in particular, are susceptible to rapidly changing weather conditions, which can be a matter of life and death. Because of the exposed nature of mountains, wind speeds can increase at short notice and be much stronger than people would usually experience in their day-to-day lives in populated areas. As clouds develop and rain or snow sets in, visibility can also become a significant issue, making navigation very difficult. And the weather can change significantly depending on where you are on the mountain and which direction it is coming from.

Unfortunately, it is not uncommon for stories to hit the headlines about people heading into the mountains ill-prepared for what the weather is about to throw at them. During Storm Ciara in February 2020, which saw heavy snowfall, thunder and lightning, wind speeds in excess of 81 miles per hour and a wind-chill factor of -20°C, four students had to be rescued from Ben Nevis. They had no specialist equipment, not even a map, and they were wearing trainers. It took twenty-two members of the Lochaber Mountain Rescue Team to save them. Team leader John Stevenson was not best pleased, as the four men not only put their own lives at risk, but the lives of his staff as well. It is absolutely vital to look at a mountain forecast before embarking on a day out to enjoy some of the most beautiful parts of the UK – and probably wise to wear appropriate footwear, too.

Rising air cools and condenses, forming cloud

Warm, moist air rises

Prevailing wind

RAIN SHADOW

Dry air sinks and warms

How Do We Get the Most Vibrant Autumn Leaves?

Autumn is one of the most noticeable and visually arresting of the seasons because of the changing colours of the leaves. As a result of the falling temperatures at that time, the chemical in the leaves that makes them green (chlorophyll) begins to break down, while other chemicals (including carotene) remain to give the leaves their yellow, red and brown colours.

The most vibrant displays of autumn leaves are evident when a dry summer is followed by an autumn with dry, sunny days and cold, but not freezing, nights. When temperatures begin to drop, the leaf produces something called an abscission layer that blocks a new supply of chlorophyll from being transferred to it from the roots of the tree via the trunk and branches. If there is lots of sunlight and low temperatures after the abscission layer forms, the remaining chlorophyll is destroyed more quickly and the autumn colours will be more vivid. At the same time, more red, orange and purple pigments are formed if the air is cool at night and there is a lot of sunshine during the day, whereas freezing conditions prevent those pigments from forming. Early frost, heavy wind or rain can also have an impact, as they can cause the leaves to fall off before they have developed the deep autumn colours that can be so spectacular.

cloudy [klou-dee]

adjective full of or overcast by clouds; having little or no sunshine

origins: before 900; Middle English *cloudi*,
Old English *clūdig* 'rocky, hilly'

CLOUDY

Clouds are one of nature's most consistently visible wonders, and they tell us a lot about what weather we can expect, from the white cotton-wool-type cumulus clouds that we see in fair weather, like the fluffy ones in a child's drawing, to the grey, overcast stratus clouds, signalling drizzle is on the way.

What's In a Name?

A classification of clouds was first introduced by Luke Howard (1772–1864), an amateur meteorologist and professional pharmacist who in 1803 wrote his *Essay on the Modification of Clouds*, which later influenced artists like Turner and Constable, as well as the poet Goethe:

CIRRUS parallel, flexuous, or diverging fibres

CUMULUS convex or conical heaps, extending upward from a horizontal base

STRATUS a widely extended, continuous, horizontal sheet

NIMBUS a cloud from which rain is falling

Types of Cloud

This original nomenclature was expanded and now forms the basis of our modern-day international naming system. There are ten main cloud types (within each there are different cloud species, and more than thirty types in total), and they can be classified by where they form in the sky:

1 **CIRRUS (Ci)** isolated cirrus clouds do not bring rain, but large amounts can indicate an approaching storm system.

2 **CIRROCUMULUS (Cc)** form when moist air at high altitudes reaches saturation, creating ice crystals – they don't tend to lead to precipitation.

3 **CIRROSTRATUS (Cs)** a precursor to rain or snow if they thicken into mid-level altostratus and eventually nimbostratus as the weather front moves closer to the observer.

4 **ALTOCUMULUS (Ac)** not always associated with a weather front but can still bring precipitation, usually in the form of virga, which is precipitation that does not reach the ground.

5 **ALTOSTRATUS (As)** can bring light rain or snow, and sometimes develop into nimbostratus.

6 **STRATOCUMULUS (Sc)** formed in slightly unstable air, they can produce very light rain or drizzle – seen at the front or tail end of worse weather.

7 **CUMULUS (Cu)** associated with fair weather, as they do not tend to lead to precipitation; however, when they grow taller, they are known as towering cumulus and can be associated with showers.

8 **CUMULONIMBUS (Cb)** develop from cumulus when air mass is highly unstable, producing thunderstorms, showers and sometimes hail.

9 **NIMBOSTRATUS (Ns)** bring constant precipitation and low visibility.

10 **STRATUS (St)** often associated with drizzle – stratus that reaches or is generated at the surface of the Earth is known as fog.

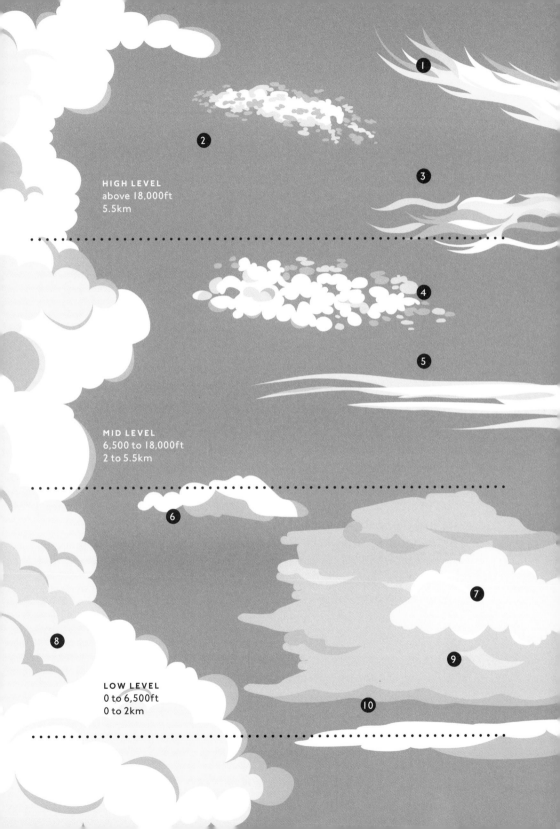

HIGH LEVEL
above 18,000ft
5.5km

MID LEVEL
6,500 to 18,000ft
2 to 5.5km

LOW LEVEL
0 to 6,500ft
0 to 2km

How to Make a Cloud

Many people believe that clouds are just made of water vapour, which is a gas, but this is not true. Water vapour is invisible and it is around us all the time in the air. Sometimes there is more water vapour in the air and it feels humid or muggy. At other times, the air has less water vapour and it feels drier and fresher.

When air rises, it cools – this is why the air tends to get colder the further you are from sea level. Cold air can't hold as much water vapour as warm air, so as the air cools it becomes saturated and the water vapour in it condenses. This means it turns from a gas to a liquid, much like when you get condensation on a cold window. When the water vapour turns to a liquid in the sky, it forms lots of tiny water droplets that cling to dust and other particles. It is this group of little water droplets suspended in the air that becomes visible as the cloud we see. These droplets of water are only about a hundredth of a millimetre in diameter, but the cloud is made up of a large collection of these. If the cloud is high up enough in the sky and the air is cold enough, the cloud is made of lots of tiny ice crystals instead, which gives it a thin, wispy appearance.

The main mechanism for cooling air is to force it to rise. The reason air cools when it rises through the atmosphere is because it encounters lower pressure. This causes the air to expand, and it's this expansion that causes the air to cool. We see this in action when we deflate the tyres on a bicycle. The air immediately cools as it escapes from the high pressure of the inner tube to the lower pressure of the surrounding environment. That's why it feels so cold if you put your hand next to the escaping air. This also explains why breathing on your hand feels colder than blowing on it. When you blow on your hand, you are forcing air through a smaller gap in your lips, making the air move from a higher- to a lower-pressure environment. The air will therefore expand and cool as it is blown out of your mouth.

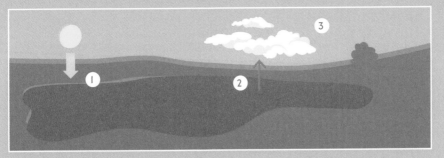

TO MAKE A CUMULUS CLOUD
1 Sun-heated ground
2 Warmed air rising into the sky
3 Fluffy peaks of cumulus

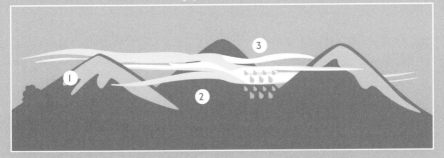

TO MAKE A STRATUS CLOUD
1 One mountain or hill
2 Air currents with nowhere to go but up the steep terrain
3 A drizzly stratus

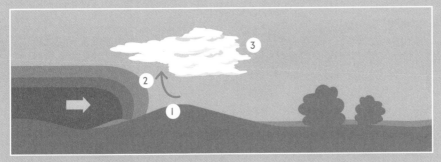

TO MAKE NIMBOSTRATUS CLOUD
1 Take a weather front, where cold and warm air meet
2 Allow the warm air to rise
3 Dark, foreboding mounds of nimbostratus

The Water Cycle

Because clouds are the source of precipitation, they are also vital to the Earth's hydrological or water cycle.

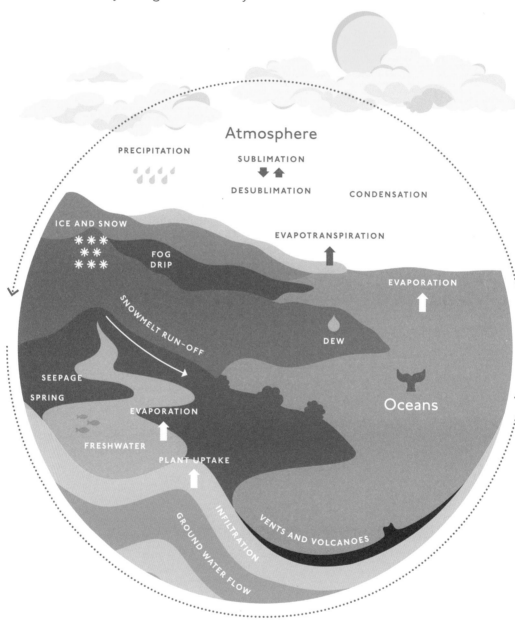

Atmosphere

PRECIPITATION

SUBLIMATION

DESUBLIMATION

CONDENSATION

ICE AND SNOW

FOG DRIP

EVAPOTRANSPIRATION

EVAPORATION

SNOWMELT RUN-OFF

DEW

SEEPAGE

SPRING

EVAPORATION

Oceans

FRESHWATER

PLANT UPTAKE

INFILTRATION

GROUND WATER FLOW

VENTS AND VOLCANOES

Why Are Clouds So Important?

Clouds are essential to the atmospheric system of the Earth in a number of ways. One of the most noteworthy, in light of the changing climate, is the way in which clouds reflect solar radiation and absorb infrared energy. This helps to create the necessary energy balance to allow life to thrive on Earth.

Why Isn't the Sky Purple?

Although light from the sun looks white, it is really made up of a spectrum of many different colours, as we can see when they are spread out in a rainbow. At one end of the spectrum is red light, which has the longest wavelength, and at the other end are blue and violet light, which have much shorter wavelengths. When the sun's light reaches the Earth's atmosphere, it is scattered, or deflected, by the tiny molecules of gas (mostly nitrogen and oxygen) in the air. Because these molecules are much smaller than the wavelength of visible light, the amount of deflection depends on the wavelength. This effect is called Rayleigh scattering, named after Lord Rayleigh (1842–1919), who first discovered the phenomenon. The shorter violet and blue wavelengths are scattered the most strongly, so more blue light is directed towards our eyes than the other colours. The sky doesn't look purple, despite the violet light being scattered even more strongly than the blue, because there isn't as much violet in sunlight to start with, and our eyes are much more sensitive to the colour blue.

Why Are Clouds White?

As light passes through a cloud, it interacts with larger water droplets, which scatter all the colours in the spectrum of light almost equally. This means that the scattered sunlight remains white, making the clouds appear white against the background of a blue sky.

How Much Does a Cloud Weigh?

While there are many factors that will determine the exact amount of water a cloud can hold – including temperature, altitude and pressure – we can work with an estimate of about 0.5 grams of water per cubic metre in an average cumulus cloud. This means an average-sized cumulus would weigh about 400,000 kilograms, roughly the same as an Airbus A380.

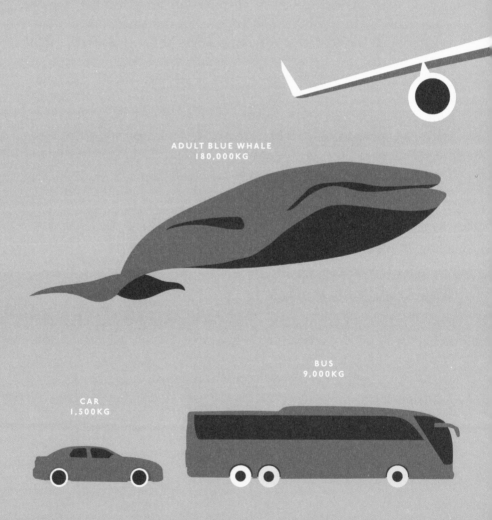

ADULT BLUE WHALE
180,000KG

BUS
9,000KG

CAR
1,500KG

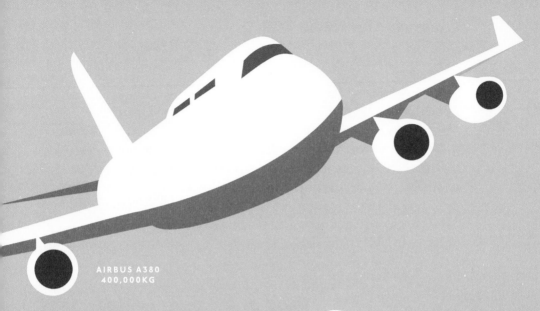

AIRBUS A380
400,000KG

AVERAGE CUMULUS
400,000KG

AFRICAN ELEPHANT
5,500KG

Clouds and Air Travel

From the vigorous updraughts and downdraughts of cumulonimbus to the poor visibility associated with low stratus, clouds can be very dangerous for aircraft. One of the biggest hazards is icing, which occurs when the temperature of microscopic water droplets within a cloud drops below 0°Celsius without freezing into ice. Unlike the ice cubes in your freezer, these so-called 'supercooled' water droplets remain liquid at temperatures below freezing, because pure water needs something solid around which to form an ice crystal. The wings and engine of an aeroplane as it flies through the sky can provide just such a solid structure. Unless the aircraft is properly anti-iced before take-off, there's a risk that supercooled water droplets will freeze instantaneously on impact with it, leading to potentially fatal malfunctions.

Fire Clouds

The Earth's surface is usually heated by the sun, but wildfires and volcanoes can also cause intense heating, which can lead to the rapid formation of clouds known as pyrocumulus. If enough water vapour is available, a pyrocumulus can become a thunder cloud, which is called a pyrocumulonimbus – these can produce dramatic lightning displays.

If the pyrocumulus cloud formed as the result of a manmade fire, it would fall into the category of cumulus homogenitus – or clouds that are created as a result of human activity. Good examples of these are the condensation trails made by aircraft, often referred to as contrails, and clouds resulting from industrial processes, such as cumuliform clouds generated by rising thermals above power-station cooling towers.

Stuck in the Clouds

The base of a cumulus cloud can be a dangerous place when thermal updrafts contribute to a phenomenon known as 'cloud suck'. In Australia in February 2007, paraglider Eva Wiśnierska-Cieślewicz was sucked up into a cumulonimbus cloud, rapidly climbing at a speed of over 44 miles per hour to an altitude of 9,940 metres – close to the altitude of an airliner. Due to a lack of oxygen, she lost consciousness but miraculously came round after about an hour to land her paraglider successfully. With temperatures of around -50°C at that altitude, she was covered in ice when she reached the ground and was bruised all over her body from the impact of hailstones while she was in the cloud. Thankfully, her injuries were superficial, and she lived to tell the tale … and to fly again. And although the phenomenon is incredibly rare, and there are no recorded occurrences in the UK, you would still be wise to watch out for cumulus clouds if you are going paragliding this weekend!

CLUES IN THE CLOUDS

There are a couple of other instances of popular weather sayings that have an element of truth about them. 'In the morning mountains, afternoon fountains' means that if clouds already appear large and mountain-like in the morning, by afternoon they'll likely have grown much larger and become cumulonimbus, which can lead to heavy downpours and possible thunderstorms. 'If in the sky you see cliffs and towers, it won't be long till there will be showers', on the other hand, relates to the fact that in unstable air, clouds have a tendency to grow taller and appear like towers. Taller clouds often lead to deeper convection and therefore make showers more likely.

Shepherd's Delight?

Clouds are an invaluable source of information when it comes to fore-casting the weather, and not just for trained meteorologists. People throughout history have used cloud formations in particular to predict the weather. One of the most famous examples from folklore is the saying 'Red sky at night, shepherd's delight. Red sky in the morning, shepherd's warning', which dates to biblical times. But is there any scientific basis to it?

The saying is most reliable when weather systems predominantly come from the west, as they do in the UK. A red sky appears when dust and small particles are trapped in the atmosphere by high pressure. This scat-ters blue light, leaving only red light to give the sky its notable appear-ance. Because high pressure moving in from the west is generally an indicator of good weather approaching, a red sky at night means that the next day will usually be dry and pleasant. A red sky in the morning tends to indicate that the high-pressure weather system has already moved east and the good weather has passed, most likely making way for a wet and windy low-pressure system later in the day. So, a red sky at night, or in the morning, can be helpful in working out what weather is to come, but it's still worth checking the Met Office forecast before you head out without a brolly (or with your sheep!).

Is that a UFO . . . or Just a Cloud?

Lens-shaped lenticular clouds (*altocumulus lenticularis*, 'like a lens') form when the air is stable and winds blow across hills and mountains from the same or similar directions at different heights through the troposphere. Although these strange, unnatural-looking clouds are rare in the UK they do sometimes form here. They look a lot like the flying saucers seen in science fiction, and they are therefore believed to be one of the most common explanations for UFO sightings around the world.

When air blows across a mountain range, in certain circumstances, it can set up a train of large standing waves in the air downstream, rather like ripples forming in a river when water flows over an obstruction. If there is enough moisture in the air, the rising motion of the wave will cause water vapour to condense, forming the unique appearance of lenticular clouds.

On the ground, they can result in very strong gusty winds in one place, with still air only a few hundred metres away. Pilots of powered aircraft tend to avoid flying near lenticular clouds because of the turbulence that accompanies them. Skilled (and brave) glider pilots, on the other hand, like them, because they can tell from the shape of the clouds where the air will be rising.

Look to the Skies

*HAMLET Do you see yonder cloud that's
almost in shape of a camel?*

POLONIUS By th'mass and 'tis, like a camel indeed.

HAMLET Methinks it is like a weasel.

POLONIUS It is backed like a weasel.

HAMLET Or like a whale.

POLONIUS Very like a whale.

(HAMLET, ACT III, SCENE II, 361–67)

Many of us will have looked upwards at the clouds on an otherwise sunny day and seen fantastic shapes: at one moment, we might see a dinosaur; at the next a dolphin … or even a camel, like Hamlet. This phenomenon is known as nephelococcygia, a form of pareidolia, which is the human tendency to interpret a stimulus as an object already known to us. It is the same reason why we see faces in strange places.

Cloud Atlas

The first photographic cloud atlas was published in 1896. It was based on the works of Hugo Hildebrand Hildebrandsson and Ralph Abercromby, who published a classification of clouds based on the work of Luke Howard. In the 1880s, Abercromby twice travelled around the globe in order to assure himself that cloud forms were the same in all parts of the world.

Many of us will have taken photographs of the clouds, looking down on them from an aeroplane window. Some of the very earliest photographs of clouds taken from above were by C.K.M. Douglas while working for the Meteorological Research Flight (MRF) during the First World War. He and the other pilots involved in the programme observed weather elements at height in order to assist the Royal Artillery, but they took the opportunity to further our understanding of the upper atmosphere and clouds at the same time.

1 Stratus
2 Cumulonimbus
3 Stratus
4 Cirrus

5 Cumulus
6 Altocumulus
lenticularis

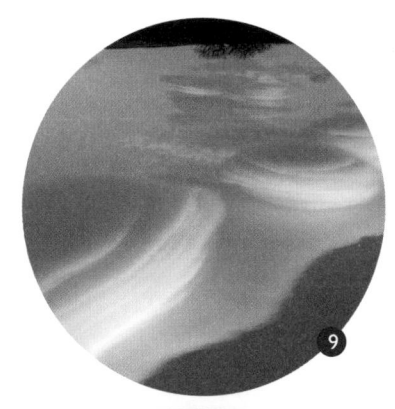

Name That Cloud

showery [shou-uh-ree]

adjective characterised by or abounding with rain showers

origin: before 950; Middle English *shour* (noun), Old English *scūr*; cognate with German *Schauer*, Old Norse *skūr*, Gothic *skūra*

SHOWERY

Also Known As...

FLIST (Scotland): a sudden shower accompanied by a squall

PILMER (England): a heavy rain shower

HASTER (England): a violent rainstorm

HAUD (Scotland): a squall

PLASH (England): a downpour of rain

PERRY (England): a squall

LAND-LASH (England): a heavy fall of rain accompanied by a high wind

GOSLING BLAST (England) / a sudden squall of rain or sleet

Most people think of showers as brief and usually light rainfall, but there are two problems with this definition: showers are not always brief, and they are not always light – as most of us have probably found out to our cost at some point.

In meteorology, we use a different, slightly less catchy, term when we think about showers: convective rain (although we can also experience convective sleet, snow and hail, so we sometimes use the term 'precipitation' to describe all the different types of frozen or liquid water that can fall through the sky from clouds). Most of the time, convective rain is a brief burst of precipitation, but sometimes it can be prolonged and heavy, resulting in flooding. And because showers often catch us unawares, they can create the sort of conditions that can spoil a day's play at Wimbledon or force you to eat your picnic in the car. We've all been there.

Don't Forget Your Brolly

The latest climate modelling suggests that we are likely to see increases in the intensity of heavy summer rainfall events in the future. It is also predicted that there will be an extension of the convective season beyond the summer months, with significant increases in heavy hourly rainfall intensity in the autumn. This means that rainfall events that typically occurred only once every two years will now occur 25 per cent more often. This will have several implications for how we manage water, with urban areas in particular likely to experience more frequent and severe surface-water flooding. Despite the intensity of hourly rainfall being projected to increase in the future, however, summers are projected to become drier overall.

What Is Convection?

Convective rain is essentially rain that is caused by convection in the atmosphere, which is the transfer of heat by the movement of either liquid or gas due to differences in the temperature in one area compared to another. For example, when you place a pot of water on a hot stove, the water is heated from the bottom. Because warm water is less dense compared to cold, bubbles rise to the top. That's also how the atmosphere works. The sun heats the ground, and the ground heats the air immediately above it. Because the ground is an uneven surface made up of a wide variety of materials, different areas warm at different rates. It's a bit like having lots of pots and pans heating at different rates on a stove. The air will therefore rise in columns, also known as thermals.

Each of these columns of air will continue to rise as long as the surrounding air is colder. On average, in the troposphere (we discuss the various layers of the atmosphere overleaf) the air gets colder with altitude. When the rate of cooling is especially high, the air becomes unstable. This could be because of much warmer air at the surface – a hot summer's day, for example – or it could be because much colder air arrives high above us, which can happen at any time of the year. When the rate of cooling is lower, the air is stable. Convective rain takes place in air that is unstable.

The larger the slice of atmosphere that is unstable and sufficiently moist, the larger the shower clouds and the heavier the showers. The air will keep rising and cooling and condensing into cloud droplets as long as the air is unstable. The top of a shower cloud occurs where the rate of cooling declines. What goes up must come down.

How Many Layers Make Up Our Atmosphere?

The atmosphere is a protective layer of gases surrounding the Earth, made up of 78 per cent nitrogen, 21 per cent oxygen and 1 per cent other gases. If the Earth were the size of an apple, the atmosphere would be as thin as an apple skin. There are four main layers, which are primarily categorised according to temperature:

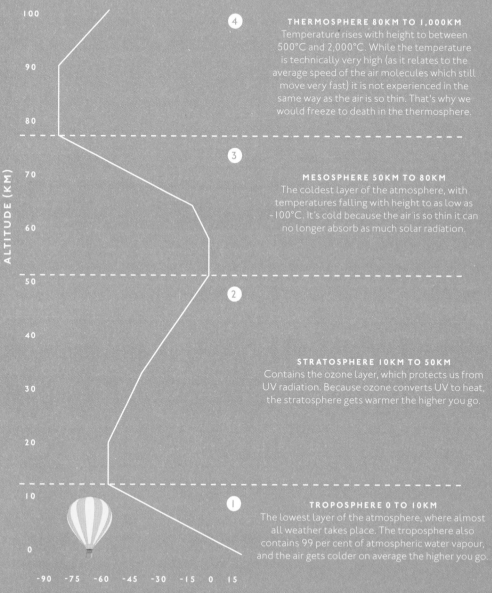

4 THERMOSPHERE 80KM TO 1,000KM
Temperature rises with height to between 500°C and 2,000°C. While the temperature is technically very high (as it relates to the average speed of the air molecules which still move very fast) it is not experienced in the same way as the air is so thin. That's why we would freeze to death in the thermosphere.

3 MESOSPHERE 50KM TO 80KM
The coldest layer of the atmosphere, with temperatures falling with height to as low as –100°C. It's cold because the air is so thin it can no longer absorb as much solar radiation.

STRATOSPHERE 10KM TO 50KM
Contains the ozone layer, which protects us from UV radiation. Because ozone converts UV to heat, the stratosphere gets warmer the higher you go.

1 TROPOSPHERE 0 TO 10KM
The lowest layer of the atmosphere, where almost all weather takes place. The troposphere also contains 99 per cent of atmospheric water vapour, and the air gets colder on average the higher you go.

ALTITUDE (KM)

100 90 80 70 60 50 40 30 20 10 0

-90 -75 -60 -45 -30 -15 0 15

TEMPERATURE

The Aeronaut

The balloon has already done for us that which no other power ever accomplished; it has gratified the desire natural to us all to view the Earth from a new aspect, and to sustain ourselves in an element hitherto the exclusive domain of birds. We have been enabled to ascend among the heavens, and to exchange conjecture for instrumental facts, recorded at elevations exceeding the highest mountains of the Earth.

JAMES GLAISHER (1809–1903)

Following the first successful flight towards the end of the eighteenth century, hot-air balloons increased in popularity throughout the rest of the nineteenth century. Although some scientific applications were considered then, it wasn't until James Glaisher came along that the value in taking measurements high up in the atmosphere was properly recognised. A founding member of the Meteorological Society and the Aeronautical Society of Great Britain, Glaisher was convinced that hot-air balloons held the key to us being able to forecast the weather.

Between 1862 and 1866, he made twenty-eight trips by hot-air balloon, usually with his co-pilot, Henry Tracey Coxwell (1819–1900). The most famous of these was on 5 September 1862 when the pair broke the world altitude record. Both men passed out as they approached approximately 8,800 metres, although it is estimated that they may have reached as high as between 9,800 metres and 10,900 metres. A pigeon they had taken on the flight died, but Coxwell managed to begin the balloon's descent before losing consciousness, and they made it safely back to Earth. Their balloon, the 'Mammoth', went on to become a popular attraction at the Crystal Palace Exhibition. And despite continuing with his dangerous pursuit, Glaisher lived to the ripe old age of 93. His exploits were most recently commemorated in the 2019 film *The Aeronauts*, starring Eddie Redmayne. Coxwell did not fare so well – his character was omitted from the film; instead actress Felicity Jones played Glaisher's fictional co-pilot. It was rather harsh treatment, as Glaisher acknowledged it was Coxwell's quick thinking that had saved them both.

SHOWERY

Ninety-nine Red Balloons or 4,300 White Radiosondes?

Nowadays, we don't need to risk lives to measure the weather in the skies above our heads. Unmanned weather balloons are launched routinely at several locations across the UK, usually a couple of times every day. Each balloon is filled with helium and carries a small radio transmitter, known as a radiosonde, and a suite of weather instruments that return readings of temperature, dew-point temperature and air pressure, while an on-board GPS allows us to infer wind speed and direction. Radiosondes reach altitudes of between 20 and 35 kilometres, depending on their size. This allows us to measure conditions in the troposphere as well as the lowest layers of the stratosphere.

Do April Showers Really Bring Forth May Flowers?

During spring, the land warms up under the strengthening sun, and by around April the land is warmer than the air. This is when showers – most commonly in the polar-maritime air mass – can develop more widely across the UK. Also, Arctic Canada and Greenland are still very cold at this time, so when the wind blows from the north-west, there is a huge contrast between the very cold air arriving above our heads and the rapidly warming ground under the strong April sunshine.

The jet stream also starts to move northwards in spring, but it's an erratic process. As the jet stream waxes and wanes in strength across the UK, the journey from winter to summer can be mixed with some warm days but also some unsettled days.

So, there is some truth in the weather lore that April showers bring forth May flowers, as April does tend to be a showery month and the warmth of May soon follows, bringing the familiar colours of the UK spring.

Shower Clouds

Convective rain falls from cumuliform clouds. These are clouds with cumulus or cumulo in their name, including cumulus, cumulonimbus and stratocumulus (*see* 'Cloudy'). *Cumulus* in Latin means 'heap', so convective or cumuliform clouds are bubbly heaps of clouds. The opposite types of cloud are stratiform clouds. These are uniform layers of clouds that can give more persistent rain not caused by convection (*see* 'Pouring').

CONVECTIVE CLOUDS

CUMULUS HUMILIS   CUMULUS MEDIOCRIS   CUMULUS CONGESTUS

The Life Cycle of a Shower Cloud

When cumulus clouds start to develop, they are shallow and tend to be wider than they are tall. These are known as cumulus humilis clouds, *humilis* meaning 'humble'. They are also known as fair-weather clouds, since they are too shallow to produce rain. As cumulus clouds become taller, they may begin to look more threatening and produce light rain. But you're unlikely to get particularly wet under these cumulus mediocris clouds; in this case *mediocris* means 'moderate'. The tallest cumulus clouds – with cloud tops up to around 6 kilometres – are called cumulus congestus, *congestus* meaning 'piled up'. When you see these in the sky, you'll want to grab an umbrella, as they are capable of producing heavy downpours. A cumulus cloud can grow from the humble humilis to the towering congestus in less than twenty minutes!

Why Do We Get So Many Showers in the UK?

Showers are common in the UK as it is subject to so many different weather influences: polar or arctic air from the north versus tropical air from the south; maritime air from the west versus continental air from the east. These contrasting conditions provide the moisture and unstable air needed for showers.

During the warmer half of the year (late spring to early autumn), warm or hot air arrives from the continent. Heating of the ground surface from the strong summer sunshine combined with this hot air at the surface imported from the continent (southerly or south-easterly winds) normally results in sunny, hot weather and some of the UK's highest temperatures. However, sufficient heat at the surface can contrast significantly with cooler air above, especially if the air at 3 to 4 kilometres above us is coming from a different direction, such as the west. This can result in higher moisture and atmospheric instability, leading to shower or thunderstorm formation. The warmer the air, the more moisture it can contain. That's why convective rainfalls in the summer can dump huge amounts of rain in a short space of time in small, localised parts of the country, resulting in flash flooding.

During the colder half of the year (late autumn through to late spring), showers are more likely to occur due to the polar-maritime air mass, which is when we get our weather from the north-west. It's called 'polar' because the air originally comes from Arctic Canada or Greenland, and 'maritime' because it has a long sea track over the Atlantic. During that journey, the air is continuously heated from below. Compared to the frigid air above, the Atlantic Ocean below is like a hot bath. It's the atmospheric equivalent of putting a pan of cold soup on the hob and turning the heat on full blast.

By the time this air arrives in the UK, it is deeply unstable, bubbling up with many heavy showers. It's still relatively cold, so the showers could be a mixture of rain, sleet, snow or hail. The Scottish mountains in particular can be plastered in snow from this set-up. During winter, when the sea is warmer than the land, most showers form over the sea and affect mainly coastal parts of the UK in the north and west. Eastern areas see fewer showers, since these areas lie in something known as a rain shadow.

Forecasting Showers

Showers are very common in the UK, but, perhaps surprisingly, there is a lot of confusion about what they are and the kind of weather people can expect. Forecasts generated by media outlets will often mention sunshine and showers or scattered showers, but does that mean it will be sunny, or does that mean it will rain? Or will it be a mixture of both? And for how long will it rain? Will the rain be heavy? What happens if it stays dry – was the forecast wrong?

Showers by their very nature are hit-and-miss – they can dump a significant amount of rain in a short space of time in some locations while others stay dry. Forecasting precisely where the heaviest showers will hit can therefore be difficult. To return to the boiling-pot analogy: we know there will be bubbles when we heat the water from below, but predicting exactly where the first, second and third bubbles will appear is very difficult. That is why expressing the forecast of showers in terms of probabilities is often the best option.

How Can You Tell If There Is Going to Be a Shower?

On a classic showery day in the UK, it can be fun to watch the clouds transform from the fair-weather cumulus that appear as temperatures rise during the morning to the towering cumulus that develop when temperatures peak in the afternoon. When these towering cauliflower-like clouds appear on the horizon, you can be sure someone is getting wet somewhere. If you have a good view and the wind is strong enough, you can sometimes tell in which direction these clouds are moving. Also, these tall clouds can block out the sun, even on the brightest of days, and strong downdraughts associated with the shower will hit the ground and spread out, resulting in a cold, gusty wind just before the rain arrives. So, if the skies suddenly darken, the wind suddenly picks up and the temperature drops, a downpour may be imminent.

The Boscastle Floods

The coastal village of Boscastle in north Cornwall was devastated by a flash flood on 16 August 2004, a once-in-400 years event. The short, steep valleys of north Cornwall – and specifically Boscastle – are particularly vulnerable to localised high-rainfall events. They collect water efficiently from the surrounding moors, and in just a few hours, channel it rapidly towards the sea.

The convective rain at Boscastle was generated by gentle south-westerly winds slightly backing over land, meaning they were turning anticlockwise. These winds came up against a sea breeze that had formed over the north coast of Cornwall following daytime heating, and a convergence line was established. A convergence line forms when winds from two different directions come together or converge. Since there is nowhere else for it to go, the air is forced upwards. This rising air can cool, condense and form shower clouds. The main difference between showers caused by rising thermals and showers produced by convergence lines is that the former will often be randomly distributed whereas the latter can form over a very specific place and remain stationary for several hours. This can result in flash flooding, just like in Boscastle.

On the day of the Boscastle flood, cumulus clouds emerging within the convergence line along the Cornish peninsula grew rapidly to around 6.5 kilometres, with small but intense rainfall cells developing continuously along this line. A single rainfall cell contains updraughts and downdraughts that act together as a single unit, and these cells can group with other cells in lines or clusters as they did at Boscastle (see 'Thundery'). This heavy rain, along with saturated soil due to the previous two weeks of wet weather and the steep river basin and catchment, all contributed to flooding. More than 200 millimetres of rain fell in around five hours into a small area at the head of the Valency river catchment above Boscastle, which was twice the average August rainfall in this area. Peak rain rates may briefly have reached as high as 400 millimetres per hour (rainfall greater than just 4 millimetres an hour is defined by the Met Office as being heavy).

WIND

RAINFALL (MM/HOUR)

	0.5 – 1
	1 – 2
	2 – 4
	4 – 8
	8 – 16
	> 32

Technology to the Rescue

At the time of the Boscastle floods, the highest-resolution computer model at the Met Office had grid squares which were 12 kilometres long – now they're 1.5 kilometres across the entire UK, meaning that we have a much more detailed picture of evolving weather conditions. Low-resolution computer models provide a vague signal, with very high intensity but localised showers averaged out over a larger area. Now, with our higher-resolution model, we are better able to predict a rainfall event like the one in Boscastle.

What Is a Rainbow and Why Do We See Them on Showery Days?

Rainbows are one of the most pleasing side-effects of convective rain (as well as being a symbol of hope!). To see one, the sun needs to be behind you, with water droplets in the air just in front of you. Falling rain droplets work best, but these water droplets can also be from fog, mist from a waterfall or even from a garden sprinkler. The sun also needs to be low in the sky, at an angle of less than 42° above the horizon. In fact, when the sun is lower in the sky, you'll see a larger rainbow arc. This also works if you see a rainbow from a position higher than the horizon – for example, from the top of a building, mountain or aeroplane. So, with water and sunlight both required for a rainbow to be seen, it makes sense that showery days are best.

In certain circumstances, it's possible to see a full-circle rainbow. That's because all rainbows are full circles, but the horizon normally obscures the bottom of the circle. This means, unfortunately, that there can't be a pot of gold at the end of a rainbow, since the rainbow is a circle and has no end. Also, rainbows do not have a fixed position in the sky. All rainbows are unique and exist only in the eye of the beholder. Imagine a straight line connecting the sun behind you through the back of your head and continuing out of your eyes in front of you. Eventually that line will reach the very middle of the rainbow's circle, which is normally below the horizon.

Rainbows form as the result of sunlight moving through higher-density droplets of water and becoming refracted. This means that the sun's white light slows down and changes direction within each water droplet. The different wavelengths, or colours, that make up the sun's white light slow by different amounts. By the time the light exits from the water droplets, it's split into its component parts and we see the seven colours of the rainbow: red, orange, yellow, green, blue, indigo and violet.

Light Refraction in a Rainbow

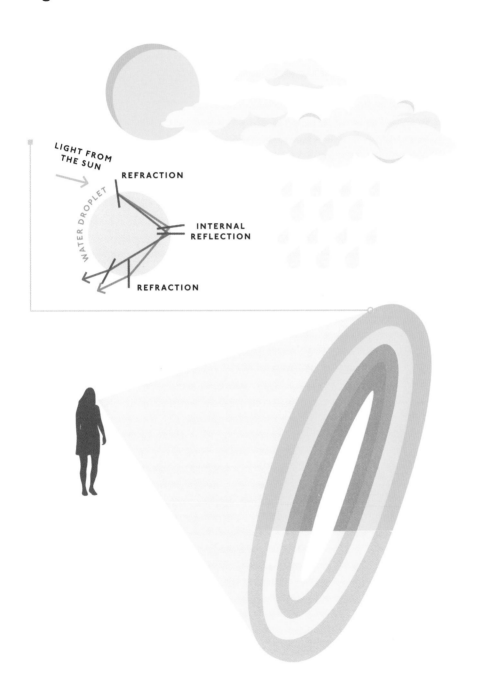

LIGHT FROM THE SUN

REFRACTION

WATER DROPLET

INTERNAL REFLECTION

REFRACTION

Four Unusual Rainbows

1 **DOUBLE RAINBOWS** occur when sunlight bounces around a raindrop not once but twice. The colours that make up the outer rainbow will be reversed compared to the colours of the inner rainbow. It's even possible to see a triple rainbow, but each additional rainbow will appear fainter than the last.

2 **MONOCHROME RAINBOWS** occur at sunset or sunrise, the sunbeam has to travel through a larger slice of atmosphere at these times of day. The blue and green components of the sun's white light are dispersed by the thicker atmosphere and the rainbow is instead dominated by reds and oranges.

3 **FIRE RAINBOWS** form when sunlight is refracted through ice crystals in high cirrus clouds; also known as upside-down rainbows, or circum-horizontal arcs – their technical name. They often appear high in the sky as quarter-circle rainbows, which is why they are sometimes referred to as 'smiles in the sky'.

4 **MOONBOWS** form at night using light from the moon instead of the sun. The physics are the same as rainbows but, since the moon is less bright and since our eyes are less sensitive to colour at night, moonbows tend to be much less colourful. They are relatively rare, since they need a bright moon and also because showers are less common at night.

The Longest-lasting Rainbow

The longest-lasting rainbow in the UK was observed in Sheffield on 14 March 1994 – it was visible for approximately 6 hours, from 9am to 3pm. This stood as the global record until it was surpassed by a rainbow in Taiwan that persisted for 8 hours and 58 minutes on 30 November 2017. Six hours may seem like a long time for a rainbow to remain visible, but there had been widespread daily rainfall totals above 10 millimetres in the Peak District just to the west of the city that day, while Sheffield, in the rain shadow, recorded 9.2 hours of sunshine.

Previously, the *Guinness World Records* gave the longest-lasting rainbow as being visible for more than four hours along the north coast of Wales on 14 August 1979, while the Fastnet storm was taking place.

Photogenic Skies

Rainbows aren't the only photogenic phenomena you might spot on a sunshine-and-showers day. Here is some other sky candy to look out for.

1 **CREPUSCULAR RAYS** can occur when light from the rising or setting sun becomes scattered in hazy conditions caused by dust, smoke and other dry particles in the atmosphere to produce sunbeams. These rays stream through gaps in clouds or between other objects. Although the beams seem to converge at a point beyond the cloud, they are actually near to parallel.

2 **VIRGA**, Latin for 'twig' or 'branch', is the streak of rain, snow or hail you might see falling from a distant cloud and evaporating before reaching the ground. If it does reach the ground without evaporating, it's called a precipitation shaft.

3 **TOWERING CUMULUS CLOUDS** will give you a soaking if you're standing underneath them, but from afar towering cumulus clouds can look particularly majestic.

4 **CLOUD FALLOUT** happens at the end of the life cycle of towering cumulus clouds, which often break up into many smaller clouds of different varieties and at different heights. Anything from stratus and stratocumulus to altocumulus can litter the skies once the showers have dispersed.

5 **STUNNING SUNSETS** occur towards the end of the day, making the skies look most interesting and varied. As the temperature falls, the development of new showers slows, while older shower clouds continue to break up. A mixture of different types of cloud at different heights, just as the sun dips below the horizon, provides ideal conditions for a beautiful sunset. It's often the medium-level clouds such as the altocumulus, at around 2 to 7 kilometres high, that are best illuminated from below by the setting sun.

6 **HALOES** are created by the tiny ice crystals within high cirrus or cirrostratus cloud interacting with light to produce a halo surrounding the sun or moon.

sunny [suhn-ee]

adjective abounding in sunshine

origins: before 900; Middle English sun, sonne, Old English sunne; cognate with German *Sonne*, Old Norse sunna, Gothic sunno; akin to Old Norse sōl, Gothic *sauil*, Latin sōl

SUNNY

When the sun shines in the UK, it is usually cause for celebration, especially because we are not guaranteed the same amount of consistently sunny weather that some other countries enjoy – it's fair to say we are more renowned for our drizzle than our sunshine. However, when the sun does make an appearance, whether that's during the height of summer or on a crisp and clear autumn day, some rays really can brighten things up.

What Is Sunshine?

At its simplest, sunlight is electromagnetic radiation emitted by the sun. It takes 8.3 minutes for this solar radiation to reach the Earth's surface, and we perceive it as visible light; in other words, the part of the light spectrum that we can see with the naked eye. The rest of the spectrum includes infrared and ultraviolet light. The light that reaches us is filtered through the Earth's atmosphere, and we experience this solar radiation as bright sunshine and heat if it is not obscured by clouds.

Life on Earth

All weather conditions influence life on Earth in one way or another, but none is more important than sunshine. Without it there would be no life, as almost all organisms rely on it for their very existence. These are just some of the vital effects of the sun on the planet and our lives, demonstrating why the forecasting of sunny weather is such an essential service:

- **PHOTOSYNTHESIS**, the process by which plant life derives its energy source.

- **FUEL**, whether directly through the growing use of solar power or indirectly by burning trees and fossil fuels.

- **VITAMIN D$_3$**, a subtype of vitamin D, helps our bodies to absorb other minerals and avoid conditions such as rickets.

- Sunlight regulates **SLEEP PATTERNS**, with adequate sleep increasingly being recognised as one of the single-most important things we can do to maintain good health.

And the elements to watch out for:

- **PREVENTING SEASONAL AFFECTIVE DISORDER (OR SAD)**, which occurs when there is reduced access to sunlight. This often occurs in the dark, grey winter months.

- **PROTECTING FROM RADIATION** – too much exposure to sunlight can lead to skin cancers and problems with our eyes. You need to know when to slap on the suncream or stay in the shade.

How Is Sunshine Measured?

For many years the only instrument for measuring sunshine duration was the Campbell-Stokes Recorder, invented in 1853 by John Francis Campbell and modified in 1879 by George Gabriel Stokes. A glass ball focuses the sun's light onto a card that burns – the length of the burn mark determines how many hours of sunshine were received at that location on that day. Most of the sunshine records we have were collected using this instrument, and it continues in use today at many non-automated climate stations. However, this method of measurement significantly overestimates sunshine duration on days when the sun is frequently shaded by passing clouds, and as such it is no longer the instrument of choice.

Nowadays, sunshine is more often measured by a special light sensor. It uses an array of differently exposed photodiodes to estimate the intensity of direct radiation from the sun. Full sunshine is defined as being sunlight that reaches an intensity of 120 watts per metre squared.

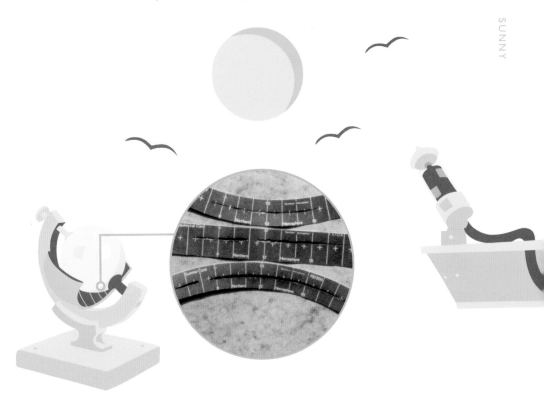

Captain Scott's Sunshine Recordings

The Terra Nova Expedition (British Antarctic Expedition) of 1910 to 1913 is best known for the famous, and ill-fated, attempt to reach the South Pole by Robert Falcon Scott and his companions. Less well known is that this intrepid British expedition also had a major scientific remit and spent far more time collecting observations and data about the Antarctic continent than in the race for the South Pole. One of the sets of data it collected was meteorological observations. Recorded when the men were living in twenty-four hours of daylight, this series of cards shows continuous sunshine for the entire period of each card, adding up to a total of 112 hours and ten minutes.

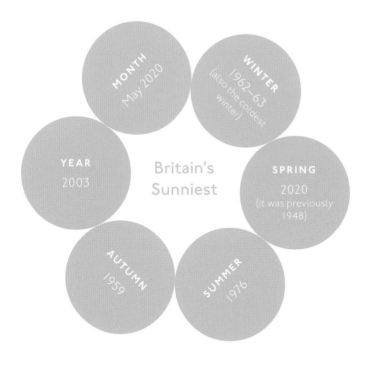

MONTH
May 2020

WINTER
1962–63
(also the coldest
winter)

YEAR
2003

Britain's
Sunniest

SPRING
2020
(it was previously
1948)

AUTUMN
1959

SUMMER
1976

When Does the Sun Shine In the UK?

Sunny days in the UK need clear, cloudless skies and stable atmospheric conditions. Days when we get a lot of sunshine are therefore often associated with areas of high pressure, which are also known as anticyclones. In anticyclonic conditions, the air descends, which can reduce cloud formation and lead to clear blue skies, light winds and settled weather.

Summer anticyclones can result in warm and sunny days, and potentially sustained dry periods – the sort that many of us wish for but which often prove to be elusive. Winter anticyclones, on the other hand, can lead to sunny but cold, dry and frosty days, perfect for long walks in the country. But high pressure alone is not enough to guarantee sunshine. Sometimes stratocumulus cloud can develop during high pressure in winter, resulting in cold and dry days with no sunshine. And areas of low pressure can lead to a sunny day interrupted by showers, which, let's face it, are what most of us probably expect when it comes to the Great British Summer.

Anticyclone Formation

With high pressure, descending air suppresses weather development…

…leading to often calm, clear or sunny conditions but also trapped cloud or fog

With low pressure, winds circulate rapidly inwards and upwards

…as air rises and cools, clouds and precipitation form

Storms are deep areas of low pressure that bring strong winds and heavy rain

We are also more likely to see sunny weather when the UK is subject to a dry, continental airflow (coming from east to west), whereas cloudy weather is more likely to be the result of a moist oceanic airflow (from west to east). And our sunniest weather in the summer is often the result of a tropical continental air mass. This is essentially a southerly wind, originating in continental Europe or North Africa, which are both relatively dry and warm locations.

Forecasting Sunshine

Forecasting sunny weather is in part about observing what is not there – namely, clouds. The same process for forecasting cloud formation and coverage can therefore be used to determine how sunny it is likely to be. We also use our knowledge of atmospheric pressure to predict sunny weather, along with satellite imagery that gives us a picture of the sky from above and helps forecasters provide greater detail about expected sunshine levels.

THE PINE CONE TEST

Most traditional, non-scientific methods for forecasting weather tend to be unreliable or grounded in myth. There is one example, though, that can be useful in forecasting fair weather: pine cones open when sunshine is on the way. The opening and closing of pine cones is dictated by humidity. In dry weather, they open out as the scales shrivel and stand out stiffly, whereas in damp conditions the increased moisture in the air allows the cones more flexibility and they return to their normal closed shape. Most of us probably don't have easy access to a supply of pine cones, though, so the Met Office weather forecast is still your best bet if you are planning a summer trip to the seaside.

Who Needs a Thermometer When You've Got a Cricket?

Did you know that the sound of crickets chirping on a warm summer evening can tell you what temperature it is? The frequency of a cricket's chirps is affected by air temperature, so if you count the number of chirps you hear over a period of 25 seconds then divide by 3 and add 4, you will arrive at an approximate temperature in Celsius.

Enjoy It While It Lasts

Because the UK is situated at the edge of the Atlantic Ocean and is subject to prevailing south-westerly winds and their associated weather fronts, we enjoy fewer days of sunshine on average than some of our European neighbours. Parts of Scotland, the north of England and Ireland see fewer than 1,200 hours of sunshine per year (an average of 3.3 hours per day), with the rest of the country not much better off at between 1,200 and 1,600 hours per year (3.3 and 4.4 hours a day), although there are, of course, local exceptions. This compares with more than 2,500 hours per year (6.8 hours) in large areas of southern Europe. This perhaps helps to explain why we Brits have a tendency to go a bit mad at the first sight of sunny weather – a sunny day is not something we can take for granted!

SUNSHINE DURATION ANNUAL AVERAGE (1981–2010)

AVERAGE VALUE (HOURS)

> 1600

1500–1600

1400–1500

1300–1400

1200–1300

1100–1200

1000–1100

<900

The Sunniest Places in the UK

The number of hours of bright sunshine is controlled by the length of day and by cloudiness. In general, December is the dullest month and May or June the sunniest. December 1956 took the biscuit for the dullest ever, with a monthly average of just 19.5 hours of sunshine throughout the UK. The sunniest places on mainland UK are along the south coast of England, with more than 1,750 hours of sunshine each year on average, while the Channel Islands enjoy more than 1,900 hours.

BRITAIN'S DULLEST MONTH
19.5 hours of sun in December 1956, throughout the UK

SUNSHINE RECORDS IN THE COUNTRIES OF THE UK

COUNTRY	HIGHEST MONTHLY TOTAL (HRS)	DATE	LOCATION
England	383.9	July 1911	Eastbourne, Sussex
Northern Ireland	298.0	June 1940	Mount Stewart, County Down
Scotland	329.1	May 1975	Tiree, Argyll & Bute
Wales	354.3	July 1955	Dale Fort, Pembrokeshire

Ask anyone about the weather where they live, and they will probably claim to live in a microclimate. And they would be correct. The fact that the UK is a small island with highly varied terrain – from islands to highlands, from town to country, and from seaside to forest – means that the UK climate could be thought of as the sum of many much smaller microclimates, and these can provide notable exceptions to the wider trends. In the month of May, for example, parts of the Western Isles (north-west of Scotland) are sunnier than the south-east of England (based on the 1981 to 2010 average). The long-term average sunshine hours for Tiree, the most westerly island in the Inner Hebrides, in May is 239 hours (7.7 hours per day) compared to 198 hours (6.4 hours per day) at Heathrow. In fact, Tiree is one of the sunniest places in the UK during the month of May, with only Shanklin on the Isle of Wight being sunnier, with 241 hours.

The amount of sunshine we saw during spring 2020 was so remarkable it 'broke' the Met Office sunshine graph: previously, the y-axis had stopped at 600 hours. The incredible 626.2 hours of sunshine recorded across the UK from 1 March to 31 May 2020 broke the previous record set in 1948 of 555.3 hours.

SUNSHINE DURATION UK SPRING
(HOURS)

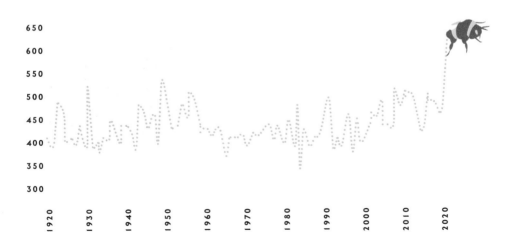

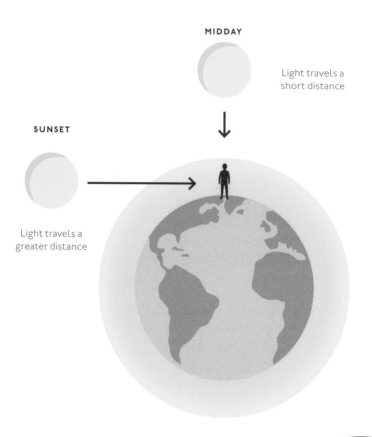

MIDDAY

Light travels a
short distance

SUNSET

Light travels a
greater distance

Why Are Sunrises and Sunsets Red?

At sunrise and sunset, the sun is very low
in the sky, which means the sunlight we
see has to travel through more of the
atmosphere before it reaches us. Because
blue light is scattered more strongly by
the atmosphere in a process known as
Rayleigh scattering (*see* 'Cloudy'), it tends to
be scattered several times and deflected away in
other directions before it gets to us. This means that there is relatively
more yellow and red light left for us to see. The results can be spectacular.

What's the Best Weather If You Suffer from Hay Fever?

Alongside the seasonal peaks and troughs in pollen levels, with the hay-fever season typically running from March to September, the weather plays an important role in the varying levels from day to day. Optimum conditions for the release of grass pollen are warm sunshine with enough breeze to carry the grains through the air. For grass, a maximum temperature between 18°C and 28°C could give a high count if it's a dry day with low humidity and a gentle breeze. Trees respond best when the temperature range is between 13°C and 15°C. However, if the temperature rises above 28°C, all pollen levels decrease. If several dry days occur in a row, then pollen supply can run out altogether. Conversely, the amount of daylight, or 'photoperiod', is also crucial to pollen production because of photosynthesis. If there is a particularly cloudy spell of weather, plants and trees produce less pollen because they are getting less light.

Wet weather temporarily suppresses pollen release, but occasional bursts of rain are necessary to prevent exhaustion of the supply. Since pollen peaks during the morning, shortly after its release, the timing of any rainfall is crucial. A wet morning will help keep levels low all day, while an afternoon shower may have less of an impact.

Weather data from the Met Office combined with expertise from the National Pollen and Aerobiological Unit and Pollen UK are used to produce pollen forecasts, ranging from low to very high, for up to five days ahead for every part of the country. While many people start sneezing only when pollen levels are high, others are sensitive even when levels are low. And with so many weather factors to take into consideration, an antihistamine may be the best idea if hay fever is preventing you from enjoying a beautiful sunny day.

POLLEN LEVEL DEFINITIONS FOR GRASS

- **LOW:** 0–29 grains/m³
- **MODERATE:** 30–49 grains/m³
- **HIGH:** 50–149 grains/m³
- **VERY HIGH:** 150 grains/m³ or more

Droughts

Although many of us long for sunny weather, there are downsides as well. Sustained periods of fair, sunny and dry weather can lead to droughts, even in the generally wet British Isles. The drought that lasted from 1975 to 1976 was the most significant for at least the last 150 years in the UK and is usually regarded as a benchmark against which all other droughts are now compared. Much of England and Wales received less than 65 per cent of its average rainfall from May 1975 to August 1976, with some parts of southern England receiving less than 55 per cent. It doesn't necessarily need to be hot to be dry, but this drought also coincided with a major heat wave (see 'Scorching') from late June to early July 1976, before abruptly ending with very wet weather in September and October 1976.

THE WORST DROUGHTS ACROSS LOWLAND ENGLAND SINCE 1910

Duration (months)
Total rainfall (mm)
Deficit (mm)
🌢 Total rainfall as % of 1981–2018 average

AUG 1920 –DEC 1921	APR 1933 –NOV 1934	FEB 1943 –JUN 1944	AUG 1947 –SEP 1949	DEC 1963 –FEB 1965	AUG 1972 –MAY 1974
17 months	20 months	17 months	26 months	15 months	22 months
630mm	829mm	662mm	1,181mm	639mm	995mm
401mm	348mm	295mm	340mm	240mm	300mm
🌢 61%	🌢 70%	🌢 69%	🌢 78%	🌢 73%	🌢 77%

Milton Keyneshenge

The sun has been venerated by cultures dating back to prehistoric times. In Britain during the Iron Age, it is thought that a number of mostly feminine deities relating to the life-giving sun were worshipped, including Sulvis and Sulevia, and the summer solstice played a significant role in religious ceremonies. Perhaps most famously, Stonehenge on Salisbury Plain is aligned so that the sun shines directly into the middle of the gigantic standing stones as the sun rises on the solstice, a spectacle that still sees thousands of people descend on the site even today. But did you know that the Milton Keynes grid system was also built with the sun in mind? The new town's central road is designed so that when the sun rises on the solstice, it shines straight down Midsummer Boulevard and reflects in the glass of the train station. If you travel further east along Midsummer Boulevard, you will find a sculpture called *Light Pyramid* that marks the exact point on the horizon where the sun will rise. Although this modern homage to the sun might suggest that there is more to Milton Keynes than meets the eye, it doesn't have quite the same impact as its counterpart at Stonehenge, a British cultural icon and World Heritage Site.

SUNNY

MAY 1975 –AUG 1976	AUG 1988 –NOV 1989	MAR 1990 –FEB 1992	APR 1995 –APR 1997	NOV 2004 –APR 2006	APR 2010 –MAR 2012
16 months	16 months	24 months	25 months	18 months	24 months
541mm	702mm	1,006mm	1,004mm	810mm	1,050mm
381mm	262mm	398mm	451mm	241mm	354mm
59%	73%	72%	69%	77%	75%

MONTH
Feb 1932

WINTER
1963–64

Britain's
Driest

SPRING
1893

YEAR
1887

AUTUMN
1922

SUMMER
1995

What Are UV Levels?

Because there can be negative effects from too much exposure to sunlight, the Met Office measures solar radiation and includes UV levels in its weather forecasts. UV refers to the ultraviolet part of the light spectrum, and it is this radiation that can be harmful. The forecast is expressed as a 'Solar UV Index', a system developed by the World Health Organization. The Met Office forecasters look at the effects of the position of the sun in the sky, cloud cover and the level of ozone gases in the stratosphere. Once gathered, this information is converted into an easy-to-understand index that ranges from 1 to 11+, determining your level of exposure to ultraviolet radiation. The UV index only rarely reaches eight in the UK, and seven may occur on exceptional days, mostly in the two weeks towards the end of June.

- UV can still be moderate or high when it is cloudy, particularly during the summer months, meaning you may need to protect yourself from sunburn when it is overcast.

- UV rays from winter sun in mountainous regions can be more damaging because UV rays are absorbed from the sun and reflected back from snow cover.

- The highest UV levels on Earth have been measured in the high Andes in South America. The world record occurred on 29 December 2003, when a UV index of 43.3 was detected at Bolivia's Licancabur volcano.

THE UV INDEX

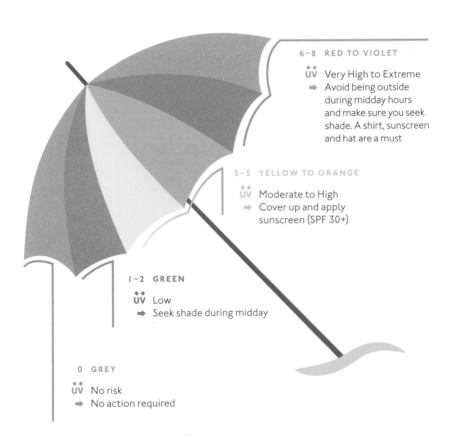

6–8 RED TO VIOLET

UV Very High to Extreme
➡ Avoid being outside during midday hours and make sure you seek shade. A shirt, sunscreen and hat are a must

3–5 YELLOW TO ORANGE

UV Moderate to High
➡ Cover up and apply sunscreen (SPF 30+)

1–2 GREEN

UV Low
➡ Seek shade during midday

0 GREY

UV No risk
➡ No action required

thundery [thuhn-duh-ree]

adjective producing thunder or a loud noise like thunder, thunder being a loud, resounding noise produced by the explosive expansion of air heated by a lightning discharge

origins: before 900; Middle English *thonder, thunder*; Old English *Thunor*; cognate with Dutch *donder*, German *Donner*; Old Norse *thōrr* 'Thor', literally, 'thunder'

THUNDERY

Thunderstorms are a series of sudden electrical discharges resulting from special atmospheric conditions. These discharges produce flashes of light and trembling sound waves, commonly known as thunder and lightning. The phenomenon is associated with convective clouds and is often accompanied by heavy rain or hail.

What Causes Thunder and Lightning?

Thunderstorms develop when the atmosphere is unstable – that is, when warm air finds itself underneath much colder air. As the warm air rises, it cools and condenses, forming small droplets of water. If there is enough instability in the air, the updraught of warm air is rapid, and the water vapour will quickly create a cumulonimbus cloud. Typically, these cumulonimbus clouds can form in less than an hour. As the warm air continues to rise, the water droplets combine to create larger droplets, which then freeze to form ice crystals. As a result of circulating air in the clouds, more water freezes on the surface of the droplet or crystal. Eventually, the droplets become too heavy to be supported by the updraughts of air and they fall as hail.

As hail moves within the cloud, it picks up a negative charge by rubbing against smaller positively charged ice crystals. A negative charge forms at the base of the cloud where the hail collects, while the lighter ice crystals remain near the top of the cloud and create a positive charge. The negative charge is attracted to the Earth's surface and other clouds and objects. When the attraction becomes too strong, the positive and negative charges come together, or discharge, to balance the difference in a flash of lightning, sometimes known as a lightning strike or bolt. The electrons shoot down from the cloud, cutting through the air at around 270,000mph. The rapid expansion and heating of air caused by the lightning produces the accompanying loud clap of thunder.

What is the Difference Between a Lightning Flash and a Lightning Strike?

Although often assumed to be the same thing, there is a key distinction between lightning flashes and lightning strikes. A lightning flash is what you can see, but this is often made up of several individual bolts of lightning, which are pulses of current that occur separately, if only hundredths of a second apart. The term 'lightning strike' refers to cloud-to-ground lightning, where lightning hits, or 'strikes', the ground.

A Bolt from the Blue

Most lightning originates from the negative part of a cloud; occasionally, however, a strike can come from the positively charged top part – this is called 'positive lightning'. When positive lightning strikes, it is forced to go around the negatively charged base of the cloud, which generally results in a more powerful lightning strike that shoots out sideways and can sometimes travel a mile or more away from the storm cloud (perhaps in blue sky) before connecting with the ground. This type of lightning strike is where we get the term 'a bolt from the blue'.

Lightning Facts

The average flash would light a **100-watt light bulb** for **three months**

Number of thunderstorms occurring at any given moment around the world: **2,000**

Number of lightning strikes every second: **100**

Number of lightning strikes a day: **8 million**

A typical house, powered by lightning alone, would need to be struck **40 times a year**

A typical lightning bolt contains **1 billion volts** and contains between **10,000** to **200,000 amperes** of current

The hottest stage of a lightning strike can reach **28,000°C**

A typical flash of forked lightning lasts for about **0.2 seconds**

THUNDERY

Types of Lightning

1 **BALL LIGHTNING**: a rare form of lightning in which a persistent and moving luminous white or coloured sphere is seen.

2 **PEARL-NECKLACE LIGHTNING**: a rare form of lightning, also termed 'chain lightning' or 'beaded lightning', in which variations of brightness along the discharge path give rise to a momentary appearance similar to pearls on a string.

3 **RIBBON LIGHTNING**: ordinary cloud-to-ground lightning that appears to be spread horizontally into a ribbon of parallel luminous streaks when a very strong wind is blowing at right angles to the observer's line of sight.

4 **FORKED LIGHTNING**: in which many luminous branches of lightning from the main discharge channel are visible.

5 **SHEET LIGHTNING**: the popular name applied to a 'cloud discharge' form of lightning in which the emitted light appears diffuse and there is an apparent absence of a main channel because of the obscuring effect of the cloud.

6 **STREAK LIGHTNING**: a lightning discharge that has a distinct main channel, often tortuous and branching; the discharge may be from cloud to ground or from cloud to air.

7 **SPRITES**: an electrical discharge that occurs high above the cumulonimbus cloud of an active thunderstorm and appears as a luminous reddish-orange, plasma-like flash.

8 **BLUE JETS**: a bright-blue discharge that projects from the top of a cumulonimbus cloud above a thunderstorm to the lowest levels of the ionosphere, 40 to 50 kilometres above the Earth.

9 **ELVES**: appear as a dim, flattened, expanding glow around 400 kilometres in diameter that last for, typically, just one millisecond. They occur in the ionosphere 100 kilometres above the ground.

THERMOSPHERE
50–85km

MESOPHERE
50–85km

STRATOSPHERE
20–50km

TROPOSPHERE
0–20km

CLOUD-TO
-CLOUD

CLOUD-TO
-CLOUD

GROUND-TO
-CLOUD

CLOUD-TO
-GROUND

FIREWORKS
150dBA

THUNDER
120dBA

Thunder

Thunder is the sound produced by lightning. When lightning strikes, the narrow channel of air through which it travels almost instantly reaches temperatures of up to 30,000°C. This intense heating causes the air to expand rapidly outwards into the cooler air surrounding it, creating a rippling shockwave that we hear as a rumbling thunderclap. Depending on its formation and location, this thunderclap can be heard as either a sudden, loud crack or a long, low rumble. Thunder lasts longer than lightning due to the time it takes for the sound to travel from different parts of the lightning channel.

The intensity and type of sound heard by the listener depends upon the conditions in the atmosphere and how close the listener is to the lightning – the louder the thunder, the closer the lightning. When a lightning strike is close, the thunder is heard as a loud clap or snapping sound. When the sound of thunder has a rumbling quality, it is the sound waves reaching the listener at different times because of the shape of the lightning strike. In rare cases, the sound of thunder at very close range has been known to cause injury to humans and damage to property.

LEAVES RUSTLING
20dBA

TALKING
60dBA

How Far Away Is a Thunderstorm?

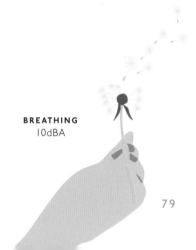

Although lightning and the associated thunderclap are generated simultaneously, thunder will always be heard after a lightning strike is seen, because light travels significantly faster than sound: the speed of light is 299,792,458 metres per second and the speed of sound 340.29 metres per second.

Your distance from a thunderstorm can be estimated by measuring the time between seeing the lightning flash and hearing the start of the thunder. The length of this interval in seconds can be divided by three to give an approximate distance in kilometres. Sometimes lightning may be seen but there is no thunder heard. This is either because thunder is rarely heard more than 20 kilometres away, or because the atmospheric conditions lead to sound bending upwards and away from the surface.

Does Lightning Strike Twice?

Contrary to the popular saying, lightning does often strike more than once. The Empire State Building in New York is struck by lightning on average twenty-three times each year and was once struck eight times in twenty-four minutes. Closer to home, The Shard in London, the tallest building in Europe, was struck multiple times during a particularly ferocious storm on 22 May 2014. Standing at 306 metres, it is the perfect conductor and is therefore likely to be hit again in the future. Thankfully, it and other tall buildings are fitted with lightning-protection systems, which channel the electricity safely to the ground.

BREATHING
10dBA

What Are the Different Types of Thunderstorm?

There are four main types of thunderstorm: single cell, multicell cluster, multicell line (or squall line) and supercell. The type of thunderstorm mainly depends on the amount of vertical wind shear between the surface and 6 kilometres in the atmosphere. Wind shear describes how wind speed and direction changes with height. So, the bigger the difference in wind velocity between the surface and 6 kilometres up, the more likely the storm will be multicell or supercell. Less than 23 miles per hour difference, taking into account speed and direction, is defined as weak shear and favours single cells, 23 to 46 miles per hour difference, or moderate shear, favours multicells, and 46 miles per hour difference and more, or strong shear, favours supercells. But there are other factors that also influence storm type (for example, available energy and humidity), so these numbers are a rough guide rather than precise definitions.

SINGLE-CELL

Single-cell thunderstorms consist of only one updraught, which initially prevents rain from falling to the ground. When the raindrops become large enough to overcome the updraught, the heavy rain shower drags colder air towards the ground. This downdraught overwhelms the updraught and subsequently cuts off the supply of warm, moist air into the storm. Typically, within 20 to 30 minutes, the storm therefore dissipates. Hail, thunder, lightning and even weak tornadoes are still possible within single-cell thunderstorms, but they don't usually cause as much or as widespread damage as those associated with supercell thunderstorms.

MULTICELL CLUSTER

Multicell clusters are made up of a group of single-cell thunderstorms in which each new cell forms on the upwind flank of the previous cell. Similar to single-cell thunderstorms, each cell has one updraught and one downdraught. However, due to moderate wind shear, the downdraught becomes separate to the updraught. As the downdraught forms (and the initial cell matures and dissipates), it travels along the surface and meets the environmental wind. These winds come together and rise into the air, forming a new updraught and a new single-cell thunderstorm. This process in which new cells are repeatedly triggered as old cells die away can continue for several hours. Although individual cells still only last for 20 to 30 minutes before moving away, the location for the mother thunderstorm can be slow-moving and result in flash flooding.

MULTICELL LINE

Thunderstorms can sometimes become organised in a line – often along or ahead of a cold front – bringing a short spell of strong wind gusts as well as heavy rain, hail and lightning. This is known as a multicell or squall line. Updraughts are found at the leading edge of a squall line, continuously feeding additional energy and moisture into the system. Strong downdraughts on the rear edge of the system, associated with heavy rain, can spread out when they hit the ground and travel ahead of a storm. This is known as the gust front, so from the perspective of someone on the ground, it's often the sudden increase in wind they encounter first, before the heavy rain or hail. Tornadoes are possible within a squall line, but it's more likely that any wind damage will be associated with the gust front.

SUPERCELL

Supercell thunderstorms, the most severe type, form in an environment of strong wind shear, normally where the winds are veering (turning clockwise) with height: for example, from the south close to the surface and from the west at 4.5 kilometres above the surface. This causes the entire storm to rotate, which results in the rotation of the storm's updraught.

- Supercells contain a single, long-lasting updraught, which rotates about a vertical axis.
- Strong wind shear induces rotation within the storm.
- A strong updraught helps to tilt the rotating column of air so that it begins to rotate around a vertical axis.

Beneath this deep rotating column of air (known as a mesocyclone), damaging tornadoes are possible. Because the updraught is kept separate from the downdraught, the storm can last for several hours.

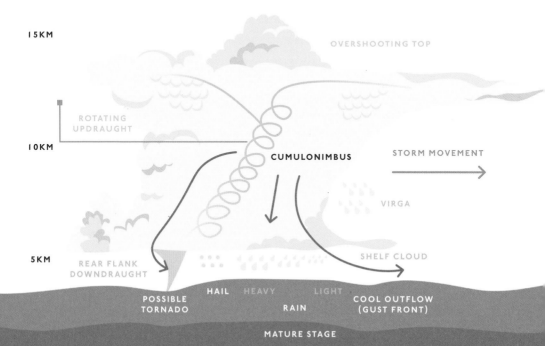

MATURE SUPERCELL THUNDERSTORM

15KM

OVERSHOOTING TOP

ROTATING
UPDRAUGHT

10KM

CUMULONIMBUS

STORM MOVEMENT

VIRGA

5KM

SHELF CLOUD

REAR FLANK
DOWNDRAUGHT

HAIL   HEAVY   LIGHT   COOL OUTFLOW
POSSIBLE   (GUST FRONT)
TORNADO   RAIN

MATURE STAGE

Supercells can produce hail, torrential rain, tornadoes, strong winds and downbursts. Downbursts are caused by particularly strong downdraughts. When the rapidly sinking air hits the ground, it spreads out in all directions, causing a damaging wind of over 150 miles per hour. The damage this causes can be mistaken for tornado damage. However, the main difference is that these winds travel in a straight line and are not rotational like a tornado. Downbursts smaller than 4 kilometres are known as microbursts. Larger downbursts are known as macrobursts.

Supercell thunderstorms are most common in Tornado Alley in the US, an area stretching from northern Texas through Oklahoma, Kansas, Nebraska, Iowa, and South Dakota. However, the first supercell thunderstorm to be defined (by Keith Browning and Frank Ludlam in 1962) occurred in Wokingham on 9 July 1959. Recent supercells in the UK have occurred on 28 June 2012 (*see* page 88) and 25 July 2019, following the hottest day on record in the UK.

LIFECYCLE OF A SINGLE-CELL THUNDERSTORM

↓ UPDRAUGHT
↑ DOWNDRAUGHTS

COLD POOL

CUMULUS STAGE   MATURE STAGE   DISSIPATING STAGE

Where Do Thunderstorms Form?

Thunderstorms are common occurrences on Earth. It is estimated that a lightning strike hits somewhere on the Earth's surface approximately 45 times per second, a total of nearly 1.4 billion lightning strikes every year.

Because thunderstorms are created by intense heating of the Earth's surface, they are most common in areas of the globe where the weather is hot and humid. Land masses, therefore, experience more storms than the oceans, and thunderstorms are also more frequent in tropical areas than the higher latitudes.

In the UK, thunderstorms are most common over the East Midlands, East Anglia and the south-east. These are also the warmest parts of the UK during the summer months, therefore they are the areas that have the highest available heat energy. In addition, we often receive imported thunderstorms from continental Europe, and the track they typically take will direct thunderstorms towards the south-east.

DAYS OF THUNDER AVERAGE (1971–2000)

AVERAGE VALUE (DAYS)

> 14

12–14

10–12

8–10

6–8

4–6

< 4

Spanish Plumes

A 'Spanish plume' is a weather event that brings an increased risk of thunderstorms as the result of warm air travelling north from the Iberian Peninsula. The storms can form over the UK, or they can move towards it, having initially developed over Spain, western France or the Bay of Biscay. There are three main ingredients typically needed for a Spanish plume:

1 Very warm air pushing north from Spain or North Africa on a southerly airflow. This can happen at almost any time of year, but during the summer months the extra warmth and moisture lead to increased energy available for storm development.

2 Cooler air at height advancing from the west associated with cold fronts.

3 Strong summer sunshine heating air at and near the surface across France and the UK.

The very warm air moving northwards from Spain towards the UK rises above the colder, less dense air already in the region. This very warm air forms a layer approximately 0.75 to 1.5 kilometres above the ground and acts as a lid, initially preventing more warm air rising from the surface. However, as the air at the surface increases in temperature, it eventually warms enough to break through this barrier and can then rise from the surface up to the higher parts of the troposphere. This air is very unstable, and the dramatic release of energy that occurs results in the formation of cumulonimbus clouds, which can in turn lead to lightning, thunder, hail, downpours of rain and sometimes gusty winds. These storms can last for several hours if there is enough energy and moisture to sustain the process and can sometimes form a large complex of thunderstorms known as a Mesoscale Convective System (MCS).

Spanish plumes most often affect southern areas of the UK, with south-eastern and southern England most at risk. These areas are closest to the source of the warm plume of air, so it is here that the contrast between warm and cool air masses is greatest. In July 2014, the UK experienced a series of Spanish plume events, as successive plumes of very warm, humid air moved north from Iberia and France and were repeatedly overrun by cooler air at higher levels in the atmosphere. The result was several outbreaks of widespread thundery activity, with the UK experiencing 62,277 recorded lightning strikes, mainly across the warmer southern half of the country, between 17 and 21 July 2014.

MEAN MAXIMUM TEMPERATURE SUMMER AVERAGE
1981–2010

**AVERAGE VALUE
(DAYS)**

> 21

20–21

19–20

17–18

16–17

15–16

14–15

< 14

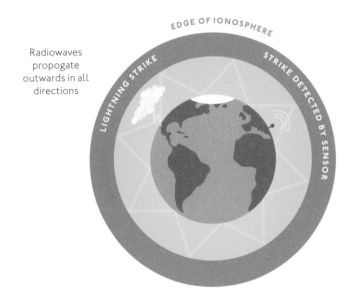

EDGE OF IONOSPHERE

Radiowaves
propogate
outwards in all
directions

LIGHTNING STRIKE

STRIKE DETECTED BY SENSOR

How Do We Detect Lightning?

Being able to detect the location of thunderstorms is of great importance, as it is not only the lightning strike that is dangerous: many other factors linked to thunderstorms can also be significant causes for concern, including intense rain and tornadoes.

When lightning strikes, it sends out pulses of radio waves, although at a much lower frequency than normal radio waves, and these can be used to detect where the thunderstorms are taking place. These pulses are known as 'sferics' and are capable of travelling great distances, because they are reflected between the surface of the Earth and a layer of the upper atmosphere called the ionosphere.

An individual sensor is able to detect a sferic, but in order to determine a thunderstorm's exact location, a network of sensors is required. When a strike occurs, a network of eleven sensors positioned around the world, known as the Met Office ATDnet system, will pick up the sferic at slightly different times, and these readings are then used to determine the exact location of the thunderstorm through a technique known as multilateration. The difference in the time taken for the sferic to reach one sensor relative to another is called the ATD (Arrival Time Difference), hence the name.

A Lightning Barrage

Perhaps the most dramatic lightning storm in recent memory was on 28 June 2012. Two lines of intense supercell thunderstorms tracked across England: one from the West Midlands to Lincolnshire, and the other from Morecambe Bay to north-east England. These brought torrential downpours, hailstones large enough to damage greenhouses and cars, and intense lightning activity – more than 64,000 strikes were recorded across the UK throughout the day, exceeding the 62,000 during the five days of the Spanish plume event two years later. To put this in context, the UK and Ireland experience on average 200,000 to 300,000 lightning strikes a year. There was intense rainfall, flash flooding, heavy hail (*see* page 93) and two tornadoes, which caused damage in the Midlands.

LIGHTNING STRIKES 28 JUNE 2012

LIGHTNING STRIKE TIMES

	0500–0700
	0600–0700
	0700–0800
	0800–0900
	0900–1000
	1000–1100
	1100–1200
	1200–1300
	1300–1400
	1400–1500
	1500–1600
	1600–1700
	1700–1800

How Likely Are You to Be Hit by Lightning in the UK?

Astraphobia is the fear of thunder and lightning. Unlike some phobias, there is some justification for being scared of lightning, as it can be deadly if it hits you. But what are the chances of actually being struck by lightning in the UK?

There is no consensus on the number of people hit by lightning, with different sources estimating the risk to be substantially different: the *British Medical Journal*, for example, suggests a one in 10 million chance, or roughly six to seven people in the UK each year, whereas the Royal Society for the Prevention of Accidents estimates that 30 to 60 people are hit each year, which approximately equates to a one in 1 million to one in 2 million chance.

The uncertainty perhaps arises from a lack of reliable records, with not every strike being verified and some going unreported. Sadly, the picture is clearer, however, when it comes to fatalities. A 2017 study in the *International Journal of Meteorology* found that 58 people were struck and killed by lightning in the UK between 1987 and 2016, or roughly two per year on average. This equates to a one in 33 million chance that you will be killed by lightning in Great Britain. By far the greatest number, at 72 per cent, were people participating in sports or leisure activities such as hill walking, fishing and golf, and 83 per cent killed were men. In addition, 26 per cent of fatalities occurred on a Sunday and 80 per cent between May and August. So, if you want to make sure to avoid the one in 33 million chance of being struck and killed by lightning in the UK, don't play golf on a Sunday in the summer if you are a man.

THUNDERY

The Derby Day Thunderstorm

For those who are not fans of horse racing, the Epsom Derby is probably most famous for suffragette Emily Davison's attempt to impede the King's horse during the 1913 running of the race. She later died from her injuries, but only two years previously the race had been rocked by another tragedy.

On 31 May 1911, a week earlier than normal in order to avoid clashing with the Coronation of King George V in early June, the Epsom Derby was hit by a severe thunderstorm. On a beautiful sunny day, with temperatures reaching 28°C, some 100,000 spectators had gathered to watch the race. As the humidity rose throughout the day, a thunderstorm moved towards the Epsom Downs, having originated to the north-west, over the Chiltern Hills. The storm struck the racecourse at around 5pm and lasted for about two hours, during which the racegoers were subjected to intense thunder and lightning and heavy downpours. In one 15-minute period from 5.30pm, it was estimated that there were 159 lightning flashes, and the spectators dispersed or took shelter wherever they could. In and around the racecourse, four people were killed and 14 injured by lightning, and four horses were killed, although none had raced that day.

Hailstorms

Hail and thunderstorms go hand in hand. Hailstones are created in convective thunderclouds (cumulonimbus) when drops of water are continuously carried up and down through the cloud and freeze. The balls of ice that result are commonly spherical or conical in shape, and their diameter can range from 5 to 50 millimetres, or even more on rare occasions, although most are smaller than 25 millimetres. Hail can only be formed in convective clouds, unlike snow, which can also be formed in weather fronts and by ascending up hills and mountains, just like rain can.

Hail is most common in western parts of Britain during the winter months. The land is cold compared to the sea at this time of year, so showers are able to form over the North Atlantic and Irish Sea, driven by rising heat energy. Once the showers reach the colder land, they tend to lose their driving force (the heat) and will therefore usually die out before they get too far inland. However, if the atmospheric conditions are right higher up in the air, the showers can sometimes continue even over cold land, and they can affect central and eastern parts of the UK too. In eastern England and south-east Scotland, hail is most frequent in spring, when temperatures are still relatively low (i.e. low enough to freeze the drop-lets of water in the cloud) but the temperature of the land is mild enough to generate shower clouds.

While Britain's most damaging hailstorms tend to occur during the summer, these are relatively infrequent. In these situations, the hot land surface has enough energy to generate very tall shower clouds, and the tops of these clouds are high enough and therefore cold enough to form hailstones. Hail in summer is most common in inland parts of northern and eastern Britain but is much less frequent in those areas during spring.

VERY BRITISH WEATHER

	INTENSITY CATEGORY	TYPICAL HAIL DIAMETER (MM)	TYPICAL DAMAGE IMPACTS
H0	Hard Hail	5	No damage
H1	Potentially Damaging	5–15	Slight general damage to plants, crops
H2	Significant	10–20	Significant damage to fruits, crops, vegetation
H3	Severe	20–30	Severe damage to fruit and crops, damage to glass and plastic structures, paint and wood scored
H4	Severe	25–40	Widespread glass damage, vehicle bodywork damaged
H5	Destructive	30–50	Wholesale destruction of glass, damage to roof tiles, significant risk of injuries
H6	Destructive	40–60	Bodywork of grounded aircraft dented, brick walls pitted
H7	Destructive	50–75	Severe roof damage, risk of serious injuries
H8	Destructive	60–90	Severest recorded in the British Isles. Severe damage to aircraft bodywork
H9	Super Hailstorms	75–100	Extensive structural damage. Risk of severe or fatal injuries to persons caught in the open.
H10	Super Hailstorms	>100	Extensive structural damage. Risk of severe or fatal injuries to persons caught in the open.

A Timeline of Some of Britain's Worst Hailstorms

MAY 1141 Wellesbourne, Warwickshire: the earliest severe British hailstorm on record, with hailstones at least 20 to 30 millimetres in diameter, although fatalities were recorded, suggesting some of the stones might have been even larger.

MAY 1697 Offley, Hertfordshire: likely Britain's most intense hailstorm, with hailstones from 110 to 140 millimetres in diameter.

SEPTEMBER 1935 Severn Estuary near Newport to Mundesley, Norfolk: the longest-track hailstorm in the UK, stretching 335 kilometres and with hailstones up to 45 to 50 millimetres in diameter.

SEPTEMBER 1958 West Wittering, West Sussex, to Maldon, Essex: the heaviest British hailstorm since official records began, with hailstones 70 to 80 millimetres in diameter and one stone weighing 190 grams, the heaviest recorded in the UK.

JUNE 2012: Thunderstorms in the West Midlands and Lincolnshire on 28 June 2012 saw some of the largest hailstones in recent years, with some measured from 50 to 90 millimetres in diameter and weighing up to 81 grams.

Ottery St Mary

Those of us who have been caught in a hailstorm know that the stones can be hard enough and travel fast enough to be quite painful if they hit you. Thankfully, however, most hailstorms are brief, and we can take shelter while they pass. That was not the case in the town of Ottery St Mary, in Devon, which was subjected to a freak hailstorm in the early hours of 30 October 2008. Witnesses reported that the size of the hailstones was unexceptional – approximately the size of a pea. What made this hailstorm such a significant event was that it lasted for roughly two hours.

The town is situated at the bottom of a valley, meaning that a huge amount of falling hail was funnelled down from higher ground and transported to the lowest parts of Ottery, leading to flash flooding and erosion of the nearby riverbanks on the way. The large amounts of hail also blocked watercourses and drainage systems, and it was therefore unable to disperse naturally. Such was the volume of hail reaching the town that the banks of hailstones were as deep as 1.2 metres in places and resembled moving ice floes drifting through the streets. Once stationary, they looked more like snowdrifts than banks of hailstones.

The unprecedented levels of hail caused significant damage, but this extreme weather event worsened as the hail gradually melted and caused further flooding, with the flood waters reaching 1.5 metres in places. By 5am, Ottery was cut off, and around 100 people had to be evacuated, with some having to be airlifted to safety. There was substantial damage to roads, housing and utility networks. It is estimated that the total cost of the clean up and repairs was about £1 million. But there is no need to worry about this happening where you live – it was a once-in-200-years event.

PICTURED: The town of Ottery St Mary, in Devon, was subjected to a freak hailstorm in the early hours of 30 October 2008.

scorching [skawr-ching]

adjective (of the weather) very hot

origins: 1400–50; Late Middle English scorchen; compare Old Norse *skorpna* 'to shrivel'

SCORCHING

Many places across the world have naturally high temperatures, but 'scorching' isn't a word traditionally associated with the British weather. We do, however, occasionally get conditions leading to very high temperatures and heatwaves – and when we do, it is often front-page news with headlines such as BRITS GO LOCO AS IT'S HOTTER THAN ACAPULCO and the GREAT BRITISH BAKE OFF.

Also Known As...

BOILING	SWELTERING
ROASTING	STICKY
BAKING	STUFFY
SCORCHIO	CLAMMY
SULTRY	STIFLING
BALMY	TROPICAL
MUGGY	SIZZLING

What Causes Very Hot Weather in the UK?

The hottest days are most frequently in summer when large anticyclones build up close to the UK and become slow-moving. In these circumstances, the jet stream may be forced to divert around the high pressure, often sending low pressure to the north of the UK and leaving much of the country dry and settled – for days or even weeks at a time. Furthermore, our highest temperatures usually occur under the influence of tropical continental air, which originates over North Africa and Southern Europe. It is most common during the summer months of June, July and August, and can lead to temperatures in excess of 30°C during the day and around 15°C to 20°C at night.

Nice Weather for Burgers

Did you know that a temperature rise from 20 °C to 24 °C can see the sales of burgers increase by more than 40 per cent?

And the first hot weekend of the year can see strawberry sales increase by 20 per cent, and barbecue meat sales by as much as 300 per cent!

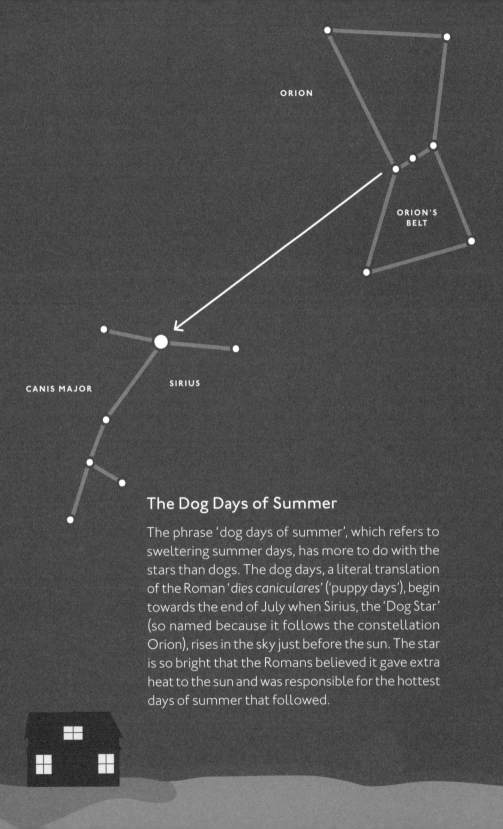

ORION

ORION'S
BELT

CANIS MAJOR

SIRIUS

The Dog Days of Summer

The phrase 'dog days of summer', which refers to
sweltering summer days, has more to do with the
stars than dogs. The dog days, a literal translation
of the Roman '*dies caniculares*' ('puppy days'), begin
towards the end of July when Sirius, the 'Dog Star'
(so named because it follows the constellation
Orion), rises in the sky just before the sun. The star
is so bright that the Romans believed it gave extra
heat to the sun and was responsible for the hottest
days of summer that followed.

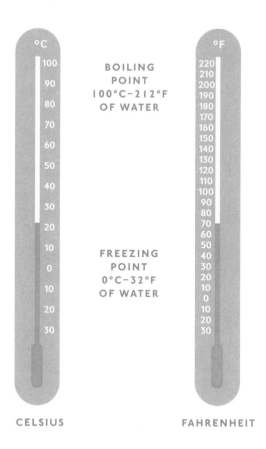

°C

100
90
80
70
60
50
40
30
20
10
0
10
20
30

BOILING
POINT
100°C–212°F
OF WATER

FREEZING
POINT
0°C–32°F
OF WATER

°F

220
210
200
190
180
170
160
150
140
130
120
110
100
90
80
70
60
50
40
30
20
10
0
10
20
30

CELSIUS

FAHRENHEIT

How Do We Measure Temperature?

Temperature is a measure of heat energy, and when measuring the weather, we usually want to know the temperature, or heat energy, of the air. To measure this, we use a thermometer, not unlike the ones you may have used at home to take your temperature when you are feeling unwell. In the past, we used mercury thermometers, invented by Daniel Gabriel Fahrenheit in Amsterdam in 1714. He also introduced the Fahrenheit scale in 1724, which along with the use of mercury led to a huge leap forward in how accurately temperature could be measured. The Fahrenheit scale is still used by many people today, perhaps most notably in the US. We use Celsius in the UK, which was invented by Swedish astronomer Anders Celsius in 1742, although it was initially referred to as the centigrade scale, only being renamed in his honour in 1948.

Mercury thermometers are no longer commonly used, as mercury is dangerous if the thermometer breaks and the liquid leaks out. Instead, we now use digital or alcohol thermometers. Like mercury, the liquid in an alcohol thermometer (usually ethanol) expands and contracts depending on the ambient temperature, giving us our reading.

What Is a Heatwave?

A heatwave is an extended period of hot weather relative to the expected conditions at that time of year. It may also be accompanied by high humidity (*see* page 104). In the UK, there is a heatwave index that sets the threshold on a county-by-county basis. If the threshold is met or exceeded at a location over a period of at least three consecutive days, a heatwave is said to be in progress.

Stevenson Screens

To measure the temperature accurately we keep our thermometers behind Stevenson screens, invented by Thomas Stevenson in 1864. Stevenson was a famous engineer who designed many Scottish lighthouses, as well as being the father of Robert Louis Stevenson, the author of *Treasure Island* and *Kidnapped*.

The Stevenson screen is a white box with slats in it to allow air to flow through. The boxes face north, which combined with their colour and the slats, give us the best measure of the current temperature, as they prevent the thermometer from getting too hot in direct sunlight.

How Often Do We Get Heatwaves?

The UK experiences occasional heatwaves but of a lesser frequency and intensity than those seen elsewhere around the globe, thanks to our more temperate climate. The years 2018, 2006, 2003 and 1976 experienced the equally warmest summers for the UK. The hottest day of the summer of 2018 was on 27 July, with 35.6°C recorded at Felsham in Suffolk.

In August 2003, the UK experienced heatwave conditions lasting ten days and resulting in a number of heat-related deaths. During this heatwave, a temperature of 38.5°C was reached at Faversham in Kent, a new high. In July 2006, similar conditions occurred, breaking records and resulting in the warmest month on record. In the summer of 2019, the 2003 maximum was then exceeded at Cambridge University Botanic Garden on 25 July, when the highest-ever temperature was recorded in the UK at 38.7°C.

HEATWAVE THRESHOLD
(DAILY MAXIMUM TEMPERATURE)

THRESHOLD
VALUE (°C)

28

27

26

25

Britain's
Hottest

MONTH
July 2006

WINTER
1988–89

YEAR
2014

WIMBLEDON
35.7°C on
1 July 2015

GLASTONBURY
31.2°C in 2017

SUMMER
1976, 2003,
2006, 2018
all tied in first place

Are Heatwaves Happening More Often?

Heatwaves are classed as extreme weather events, and therefore outside the norm, but research shows that climate change means we are seeing them more often. A scientific study by the Met Office into the 2018 summer heatwave showed that the likelihood of the UK experiencing a summer as hot or hotter than that one is a little over one in ten, and we are thirty times more likely to experience a summer as hot as 2018 now than before the Industrial Revolution because of the higher concentration of carbon dioxide (a greenhouse gas) in the atmosphere. As greenhouse-gas concentrations increase, heatwaves of similar intensity are projected to become even more common, perhaps occurring as regularly as every other year by the 2050s. The Earth's surface temperature has risen by 1°C since the pre-industrial period (1850–1900), and UK temperatures have risen by a similar amount.

Why Is It Hotter in the City?

One of the most significant microclimate effects we see in the UK is the warming of urban areas. This is particularly pronounced during clear and calm nights in the winter, and on sunny, hot days in the summer – for example, temperatures in the centre of London can be more than five degrees higher on these occasions compared to the suburbs. This is known as an urban heat island and was an effect first noted by Luke Howard (cloud classification originator). It's caused by:

- The release of heat and reflection of solar radiation by buildings.

- The absorption of heat by buildings and tarmac during the day, and its slow release overnight.

- Pollutants trapping heat above a city.

- Concrete and tarmac retaining less water than forests or fields, and therefore less of the available heat energy being used up in evaporating water, meaning more is available to heat the air.

What Is Humidity?

Hot weather can also be accompanied by high levels of humidity and the kind of oppressive, muggy conditions that make life uncomfortable for us Brits who are not used to it.

As we've seen, water can be solid (ice), liquid (water) or gas (vapour) – atmospheric humidity is a measure of water held in the air as a gas. The vapour component makes up about 99 per cent of all water held in the atmosphere. Warmer air can carry more water vapour than cooler air, if there is plenty of water available. This is because it has more heat energy to evaporate water into vapour and keep it in this state. The tropics are very warm and very humid, because the air there contains lots of water vapour. Conversely, there is very little vapour over the very cold Arctic and Antarctic, and some very warm regions are also very dry (e.g. the deserts of the Sahara), because there is very little available water to evaporate into vapour, and at about 30 degrees north or south of the equator the air descends from above and does not contain much water either.

Relative Humidity (RH) is the most common scale used to show how humid it is. It measures how close the air is to being saturated – that is, how much water vapour there is in the air compared to how much there could be at that temperature. If the RH of the air is 100 per cent, it is fully saturated.

How Do We Measure Humidity?

We measure humidity using what is known as the wet-bulb temperature, which is measured similarly to normal air temperature, but the end of the thermometer is wrapped in a damp cloth. It is then housed alongside a standard thermometer behind a Stevenson screen. Energy is needed to evaporate the water from the cloth, and this causes the air around it to cool a little (from the loss of energy), which means the thermometer usually measures a lower temperature. Using the temperatures from the two thermometers we can then calculate the humidity.

Humidity above the surface of the Earth can be measured using instruments suspended from a helium balloon called a radiosonde that is allowed to float up through the atmosphere (*see* 'Showery'). Satellites can also be used to work out the amount of water vapour in the air, or the RH. And more recently, GPS signals have been used to infer atmospheric humidity, as signal timing is affected by water-vapour quantity.

BRITAIN'S HIGHEST
TEMPERATURE
38.7°C on 25 July 2019,
at Cambridge University
Botanic Garden

Why Do We Find High Humidity So Uncomfortable?

Humidity can influence human health because it affects our thermal comfort – in other words, whether we feel too hot or too cold. When the weather is warm and humidity is high, the body finds it difficult to keep cool, because it's harder to remove heat via evaporation of sweat into the saturated air. This can lead to dangerous levels of overheating, which can be a particular problem at sporting events held during the summer months, and extra care has to be taken to ensure athletes and spectators are well hydrated and rested as often as possible.

Humans are most comfortable when the RH level is within the range of 30 to 70 per cent, but the ideal is between 50 and 60 per cent, depending on the temperature – the higher the temperature, the less humid we prefer it to be. The air can also be too dry, potentially leading to cracked skin and nosebleeds when the RH drops below 30 per cent.

After an extended period of high humidity, the air develops a 'heavy' feel. When it rains, because the air can no longer hold the vapour, humidity decreases and the air feels 'lighter'. This has been shown to lead to widespread feelings of elation. There is some research that suggests that this change in humidity, mostly in parts of the tropics (where it is humid throughout the year), may even contribute to an improvement in physical and mental health.

THE EARTH'S
WARMEST
YEARS
2015, 2016, 2017,
2018 and 2019

What Is the Greenhouse Effect?

Greenhouse gases in the atmosphere act like a blanket around the Earth. When sunlight (short-wave radiation) hits this blanket, it passes straight through and continues until it reaches the surface of the planet. The Earth then absorbs this sunlight and emits infrared radiation back out to space. As it leaves the atmosphere, the infrared radiation also hits the greenhouse gas blanket. Most of it goes straight through, but some of it is absorbed and redirected back to Earth. This traps the infrared radiation and causes the surface to heat – a process we call the 'greenhouse effect'.

6

Increasing greenhouse gas concentrations in the atmosphere increases the amount of heat retained, causing the atmosphere and Earth's surface to heat up. This is the ENHANCED greenhouse effect – commonly known as Global Warming.

Some of the sun's rays are reflected by the atmosphere and the surface of the Earth into space

Infared radiation (IR) is given off by the Earth – most escapes into space, allowing the Earth to cool …

2

4

5

Sunlight passes through the atmosphere

1

… but some IR is trapped by greenhouse gases in the atmosphere, keeping the Earth warm enough to support life

ATMOSPHERE

3

Some of the sun's rays are absorbed by the Earth's surface, causing it to warm

The greenhouse gases include water vapour (H_2O), carbon dioxide (CO_2), methane (CH_4), nitrous oxide (N_2O) and ozone (O_3). Of these, water vapour is the most abundant, but we don't worry about it as a greenhouse gas to the same extent as carbon dioxide and methane because humans are not emitting water vapour. However, in a warmer world, there will be more water vapour in the atmosphere, which will act as a positive feedback for global warming.

The greenhouse effect is critical to life on Earth. Without a blanket of greenhouse gases trapping in heat, the temperature would be bitterly cold, and humans would be unable to survive. The average temperature of the Earth at the surface is 14.5°C, but without any greenhouse gases it would be around -18°C. However, by adding extra greenhouse gases, such as carbon dioxide into the atmosphere, humans have created an enhanced greenhouse effect. The greenhouse gas blanket is now thicker and is absorbing more infrared radiation than before. In other words, the greenhouse effect is stronger, and instead of keeping the Earth at a stable temperature, it is causing the planet to heat up. This is a major factor in climate change.

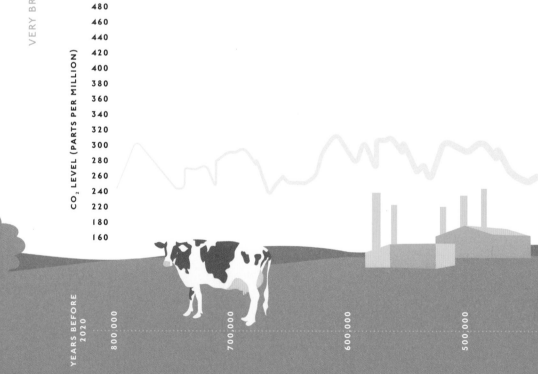

CO₂ LEVEL (PARTS PER MILLION)

480
460
440
420
400
380
360
340
320
300
280
260
240
220
180
160

YEARS BEFORE 2020

800,000   700,000   600,000   500,000

How Do We Know Climate Change Is Manmade?

In the 11,000 years before the Industrial Revolution, the average temperature across the world was relatively stable at around 14°C. During the rapid transition to industrialisation in the nineteenth century, humans began to burn fossil fuels such as coal, oil and gas for fuel, which produced energy but also released greenhouse gases such as carbon dioxide, methane and nitrous monoxide into the air.

Over time, large quantities of these gases have built up in the atmosphere. For example, during the twentieth and twenty-first centuries the level of carbon dioxide in the atmosphere rose by 40 per cent and is now more than 400 parts per million. **This level of carbon dioxide is higher than at any time in the past 800,000 years.**

Since the 1850s, the average temperature of the planet has risen by more than 1°C. This is a rapid change in terms of our global climate system. Previously, natural global changes are understood to have happened over much longer periods of time. It is also important to remember that the world is not warming evenly, so the temperature increase is higher than 1°C in some countries. The Earth's five warmest years on record were 2015, 2016, 2017, 2018 and 2019. And all of the top-ten warmest years for the UK since 1884 have occurred since 2002.

300,000   200,000   100,000   0

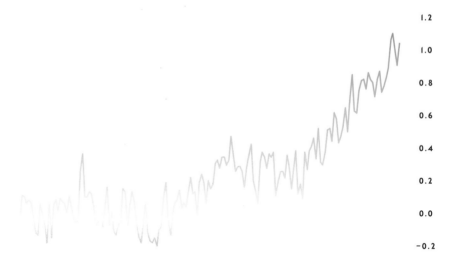

1.2

1.0

0.8

0.6

0.4

0.2

0.0

−0.2

* Compared to 1850–1900 'pre-industrial' levels. Data source – HacCRUT

How Will Climate Change Affect the UK?

UK winters are projected to become warmer and wetter on average as a result of climate change, although cold or dry winters will still occur sometimes. Summers are projected to become hotter and are more likely to be drier, although wetter summers are also possible.

Even if we do reduce greenhouse-gas emissions, sea levels around the UK will keep rising beyond 2100.* Parts of the UK will be in danger of flooding, with low-lying and coastal cities at particular risk. Heavy rainfall is also more likely. Since 1998, the UK has seen seven of the ten wettest years on record. The winter storms in 2015 were at least 40 per cent more likely because of climate change.

Farming in the UK will be affected by climate change, too. Hotter weather and higher levels of carbon dioxide could make growing some crops easier, or even allow us to produce new ones. However, with more droughts expected, water might not be as easy to access, making it harder for farmers to plan the growing season.

Floods, storms and extreme heat can cause damage to buildings, disrupt transport and affect health, so buildings and infrastructure will also need to be adapted to cope with the new conditions, and businesses will have to work around the changing climate. The government can play a key role as well, as can each individual in the choices that they make. Although it may be too late to reverse some of the effects of climate change, we can still limit the impact, but only if the UK and the rest of the world acts now.

Wildfires

Devastating wildfires have often made the news in recent years in places such as Australia, Canada, California and Greece. But the UK has also seen a recent spike in wildfires – in 2019, there were more wildfires in Great Britain than any previous year on record. Researchers have modelled the impacts of climate change on the Peak District and found an increase in wildfires there as a result of the combination of warmer, wetter winters (more vege-tation to fuel the fires) and hotter, drier summers (higher risk of ignition). There is every chance this trend will continue across the UK in the future.

* This is because the ocean and ice-covered parts of the climate will continue to respond to the carbon dioxide that we have already emitted for many years. Essentially, the CO2 continues to provide positive radiative forcing for centuries (because they have a very long atmospheric lifetime) and the system as a whole continues to take up heat long after surface temperature change has stabilised. About 90 per cent of the system heat uptake happens in the oceans, which expand under the warming and raised sea level. The glaciers and ice sheets take many centuries to adjust to the elevated surface temperature (compared to pre-industrial temperatures) and we therefore expect to see continued sea-level rise associated with ice melt of the ice sheets for many centuries and even millennia.

The Saddleworth Moor Wildfire

Most wildfires in the UK occur in remote areas, so fortunately they do not cause much damage or risk to life. However, one of the biggest wildfires in the UK in recent years was so large that the smoke from it could be seen from space. During the scorching summer of 2018, a fire broke out on Saddleworth Moor between Sheffield and Manchester. A major incident was declared on 26 June, and 150 people were evacuated from their homes as the fire began to burn out of control. This is thought to be the first time a wildfire has forced a large-scale evacuation to take place in the UK. By the end of 27 June, the blaze had been tackled by:

2 helicopters

More than
65,000
gallons of water

More than **100**
firefighters

29
fire engines

Three weeks later, on 18 July, the fire was finally extinguished – 18 square kilometres of moorland had been burned.

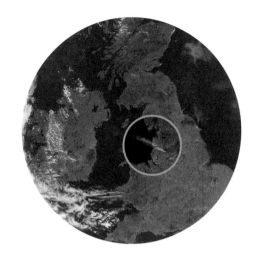

Fire Whirls and Tornadoes

During the long, hot summer of 2018, a fire whirl or fire devil – sometimes incorrectly nicknamed a fire tornado or 'firenado' – was filmed in Derbyshire. This wasn't associated with a wildfire but with a plastics factory that was ablaze.

Fire whirls are dangerous and terrifying, and thankfully, very uncommon – they are especially rare in the UK. Fire whirls occur when rising columns of hot air within a fire become disturbed and start rotating, and they are indicative of the fire creating its own wind system, which can make the behaviour and spread of a wildfire, for example, dangerously unpredictable. Normally, fire whirls are 10 to 50 metres tall, a few metres wide and last a few minutes. Temperatures within them can exceed 1,000°C. The fire whirl that occurred at the plastics factory reached a height of more than 15 metres.

Fire tornadoes, which are distinct from fire whirls, are also incredibly rare. The first confirmed case was documented during the 2003 Canberra bushfire, with wind speeds calculated to be 162 miles per hour. Fire tornadoes are spawned from pyrocumulonimbus clouds (*see* 'Cloudy'). These are cumulonimbus clouds that develop above wildfires or volcanoes following the huge amount of hot air rising and cooling vertically into the atmosphere. If this storm cloud starts rotating, fire tornadoes can result – just the same as the formation of a non-fire-related tornado (*see* 'Freaky').

blustery [bluhs-ter-ee]

adjective characterised by or subject to strong winds

origin: 1520–30; perhaps Low German *blustern*, *blüstern* 'to blow violently'; compare Old Norse *blāstr* 'blowing, hissing'

BLUSTERY

Also Known As...

What Is Wind?

To understand what makes the wind blow, we first need to understand what atmospheric pressure is. Pressure at the Earth's surface is a measure of the 'weight' of air pressing down on it. The greater the mass of air above us, the higher the pressure, and vice versa. The importance of this is that air at the surface will always move from high to low pressure to equalise the difference, which is what we know as wind.

So, wind is caused by differences in atmospheric pressure, but why do we get these differences? It's down to the rising and sinking of air in the atmosphere – cold air sinks, while warm air rises. Where air is rising, we see lower pressure at the Earth's surface, and where it's sinking, we see higher pressure. In fact, if it weren't for this rising and sinking motion in the atmosphere, not only would we have no wind, but we'd also have no weather.

How Do We Measure Wind?

The instruments used to measure wind are known as anemometers, and they can record wind speed, wind direction and the strength of gusts. In the UK, winds are measured in knots (nautical miles per hour). However, forecast winds are often given in miles per hour (where one knot is equivalent to 1.15 miles per hour) or in terms of the Beaufort Scale (*see* 'Stormy'). Wind direction is measured relative to true north (not magnetic north) and is reported from where the wind is blowing – for example, an easterly wind blows from east to west.

Wind speed normally increases the higher you are above the Earth's surface and is heavily affected by factors such as the roughness of the ground and the presence of buildings, trees and other obstacles. The optimal exposure for the measurement of wind is therefore over level ground of uniform roughness with no large obstacles within 300 metres of the tower where the instruments are housed.

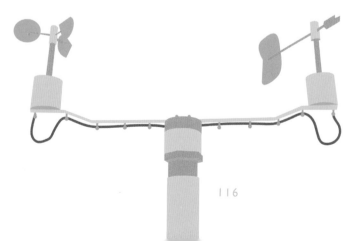

CUP ANEMOMETER

Wind speed is normally measured by a cup anemometer consisting of three or four cups, conical or hemispherical in shape, mounted symmetrically about a vertical spindle. The wind blowing into the cups causes the spindle to rotate. In standard instruments, the design of the cups is such that the rate of rotation is proportional to the speed of the wind. At intervals of no longer than five years, anemometers are calibrated in wind tunnels to ensure they remain accurate.

MEASURING WIND DIRECTION

Wind direction is measured by a vane consisting of a thin, horizontal arm carrying a flat, vertical plate at one end with its edge to the wind and at the other end a balance weight that also serves as a pointer. The arm is carried on a vertical spindle mounted on bearings that allow it to turn freely in the wind. The anemometer and wind vane are each attached to a horizontal supporting arm at the top of a 10-metre mast.

SONIC ANEMOMETER

Where wind measurements are made in extreme weather conditions, such as on the top of mountains, a heated sonic anemometer is used, which has no moving parts. The instrument measures the speed of acoustic signals transmitted between two transducers located at the end of thin arms. Measurements from two pairs of transducers can be combined to yield an estimate of wind speed and direction.

There can be rapid variations in the speed of the wind – these are referred to as gusts. Gusts are stronger inland, where the topography can be extremely varied, than over the relatively flat seas or windward coasts. Typically, gusts can be 60 per cent higher than the mean speed of wind inland, although in the middle of cities this can reach 100 per cent as a result of buildings creating an even more contrasting landscape and therefore more friction and channelling of the wind. Because wind is an element that varies rapidly over very short periods of time, it is sampled at high frequency (every 0.25 seconds) to capture the intensity of gusts, or short peaks in speed, which inflict the greatest damage in storms. The gust speed and direction are defined by the maximum three-second average wind speed occurring in any period. A better measure of the overall wind intensity is defined by the average speed and direction over the 10-minute period leading up to the reporting time. A gale is defined as a surface wind of mean speed of 34 to 40 knots, averaged over a period of 10 minutes.

BRITAIN'S STRONGEST GUST SPEED

173 mph on 20 March 1986 at the summit of Cairngorm

What Is the Difference Between Small-scale and Large-scale Winds?

Winds are generally classed as being either small- or large-scale depending on whether they have a local or global impact. Small-scale winds, such as sea breezes (see page 123), also tend to arise because of conditions in a specific area, whereas large-scale winds, including the jet streams, primarily arise because the sun's rays reach the Earth's surface in the polar regions at a much more slanted angle than at the equatorial regions. This sets up a temperature difference between the hot equator and cold poles: heated air rises at the equator, leading to low pressure, while cold air sinks above the poles, leading to high pressure. This pressure difference creates global wind circulation, as the cold polar air tries to move southwards or northwards to replace the rising tropical air.

Another important factor is the Coriolis effect from the Earth's rotation. This means that air does not flow directly from high to low pressure – instead, it is deflected to the right in the northern hemisphere and to the left in the southern hemisphere. This effect contributes to our prevailing west to south-westerly winds across the UK.

HIGHEST RECORDED WIND SPEEDS IN THE UK

COUNTRY	SPEED (MPH)	DATE	LOCATION
Scotland	142	13 Feb 1989	Fraserburgh, Aberdeenshire
Northern Ireland	124	12 Jan 1974	Kilkeel
Wales	124	28 Oct 1989	Rhoose, Vale of Glamorgan
England	118	15 Dec 1979	Gwennap Head, Cornwall

The Windiest Places in the UK

In general, the windiest parts of the UK are the north and west. This is because the prevailing west to south-westerly winds across the UK lead to northern and western areas being typically more exposed to Atlantic weather systems than the south and east. The strongest gust speed ever recorded in the UK was 173 miles per hour on 20 March 1986; however, this was at the summit of Cairngorm at 1,245 metres above sea level. The strongest gusts in each of the four nations were recorded at altitudes more representative of the environments that most of us live in.

THE TOP TEN WINDIEST AREAS OF THE UK, BASED ON THE ANNUAL AVERAGE WIND SPEED FROM 1981 TO 2010 ARE:

LOCATION	AVERAGE WIND SPEED (KNOTS)
Shetland	14.7
Buteshire	14.5
Orkney Area	14.3
Caernarvonshire	12.9
Western Isles	12.6
Argyllshire	12.1
Anglesey	11.9
Inverness	11.8
Peebleshire	11.3
Ross and Cromarty	11.3

The Windy Heights of Air Travel

Because the weather can be so unreliable in the UK, many of us choose to jet off to warmer climes for our summer holidays, or in search of the perfect ski conditions during the winter months. And although many weather types can have an effect on air travel, perhaps the most obvious when you are a passenger is turbulence – or, more accurately, clear-air turbulence (CAT). This is primarily caused when invisible air masses travelling at different speeds meet, causing disruption to the flight of the aircraft. This kind of turbulence occurs most frequently in the upper regions of the troposphere at altitudes between 7,000 and 12,000 metres, and in the vicinity of the rapid movement of air caused by the jet streams. Although the presence of cirrus clouds at heights not normally associated with cloud formation can be an indicator of CAT, for the most part it is very difficult for pilots to detect and avoid it. This is why you should always wear your seatbelt when advised.

Jet streams are also important to aviation in other ways – for example, airlines flying east will try to catch the Transatlantic jet stream to reduce journey times and lower fuel consumption, whereas flights going west will try to avoid it. However, if you arrive too early, you'll just end up circling and waiting to land. And if the jet stream is weak, this can cause delays. Flight planning is, therefore, quite a skill. That is why Met Office meteorologists work in one of only two centres in the world that produce weather charts for global aviation, helping the airlines to fly in the safest and most efficient way possible.

What Is the Foehn Effect?

In simple terms, the foehn effect is a change from wet and cold conditions on the windward side of a mountain to warmer and drier conditions on the leeward side. When air meets an obstacle such as a mountain, it is forced to rise on the windward side. As the air rises, it cools and expands due to the decreased pressure associated with an increase in altitude. Colder air holds less water vapour, so the moisture condenses to form clouds, and it rains on the windward side. The change of state from vapour to liquid water creates heat energy, and the air descends on the other side of the mountain as a pleasantly warm wind on the downward slope. Less cloud and therefore more sunshine also leads to higher temperatures on the leeward side as compared to the windward side.

In the UK, the 'Helm' wind of the English Pennines is perhaps the best-known example of the foehn effect, as it is the only one with its own name, but the most notable foehn events tend to occur across the Scottish Highlands, where the moist prevailing westerly winds encounter high ground along Scotland's west coast. This results in a marked contrast in weather conditions across the country, with the west being subjected to more wet weather, while the lower-lying east more often enjoys the warmth and sunshine of the foehn effect.

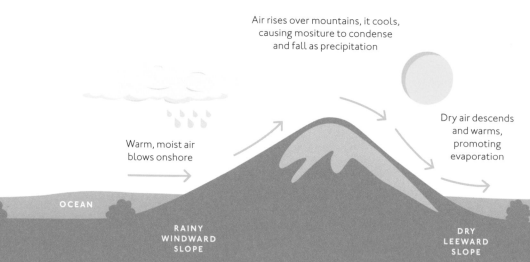

Air rises over mountains, it cools, causing mositure to condense and fall as precipitation

Warm, moist air blows onshore

Dry air descends and warms, promoting evaporation

OCEAN

RAINY WINDWARD SLOPE

DRY LEEWARD SLOPE

Why Is It Sometimes Cooler at the Beach?

Even on a hot, sunny day, it can feel cooler at the beach than further inland because of a local wind called a sea breeze. During the summer, the sun's rays heat up the ground quickly. By contrast, the surface of the sea has a greater capacity to absorb the sun's rays and is more difficult to heat – this leads to a temperature contrast between the warm land and the cooler sea. As the land heats up, it also warms the air above it. The warmer air becomes less dense and begins to rise, leading to lower pressure over the land, whereas the air over the sea remains cooler and denser, so pressure is higher there than inland. As a result of this contrast in pressure, the air moves inland from the sea to try to equalise the difference, resulting in a cooling sea breeze and giving us another good reason to head to the coast when the weather is nice.

Weathercocks

Although the weather vanes we use at the Met Office to measure wind direction are now purely functional in design, there is a long history of ornate versions adorning some of our tallest buildings. In the past, these were most often on church spires. Many of these vanes were in the shape of a cockerel, which has led to them sometimes being called weathercocks – but why a cockerel? There is no clear answer, but it may have something to do with Pope Gregory I saying in the sixth century that the cockerel, the emblem of St Peter, who according to the Bible denied Jesus three times before the cock crowed, was 'the most suitable emblem of Christianity'. As a result, the use of weathercocks became more widespread, until the ninth century when Pope Nicholas I ordered that every church steeple have one. In the Bayeux Tapestry of the 1070s, a weathercock can be seen on Westminster Abbey. These early wind vanes, although functional, were mainly for ornamental purposes, and some beautiful examples can still be seen on our buildings today.

The Most Important Weather Forecast

Probably the most important weather forecast the Met Office has ever done was in advance of the D-Day landings towards the end of the Second World War. In order for Operation Overlord, the Allied invasion of mainland Europe, to be a success, favourable weather conditions were key. These included sufficient moonlight to allow the invasion to take place in the cover of darkness but with enough light once the troops were on the ground, less than 30 per cent cloud cover to ensure good visibility and a low tide so that the landing vessels would be able to avoid obstacles placed on the beaches by the German forces. Two possible windows of opportunity when the lunar and tidal conditions would be suitable were identified: 5 to 7 June and 19 to 21 June. If these windows were missed, the invasion would have to be delayed until July, allowing the German Armed Forces time to better prepare themselves for the attack. Accurate forecasts would therefore be essential in determining whether the invasion could go ahead and when.

Meteorologists from the Army, Navy and Air Force were involved in making the observations and forecasts for the Allied forces, led by Group Captain James Stagg of the RAF and Met Office. An Allied observations chart from 1800 hours on 3 June was plotted just an hour before Stagg was due to brief General Eisenhower, the Supreme Allied Commander in Europe, about the predicted weather on 5 June. The forecast showed that the conditions would be extremely unfavourable: a low-pressure system was forecast over France, which would lead to strong south-westerly winds across the Channel, making the sea too rough for the small landing vessels. The cloud and rain associated with the cold front would also be problematic for the airborne part of the invasion. However, Stagg and his team predicted that a ridge of high pressure would follow the cold front, so Eisenhower chose to delay the invasion for twenty-four hours to 6 June.

PICTURED: D-Day Landing Weather Chart

The German forecasters, meanwhile, were at a distinct disadvantage to their Allied counterparts. The cracking of the Enigma code meant that the Allied meteorologists had access to the German weather observations as well as their own. The gap in the German information led them to believe that the bad weather would continue until the middle of June, and as such the German military commanders decided to delay sending troop reinforcements to defend themselves from the inevitable attack.

An observation chart from 0100 hours on 6 June shows that the cold front had passed, as forecast, and the weather window that had been predicted had materialised. By 1300 hours on 6 June, seaborne troops had been landing for seven to eight hours, and the invasion was a success. Not only had the forecasters got it right, but they had identified the only moment during June that would have been suitable. The weather on 19 to 21 June was bad, with Force 6 to 7 winds and gusts up to Force 8, which would have made landing impossible. Stagg sent a memorandum on 21 June highlighting this fact, on which Eisenhower wrote:

♦ Thanks – and thank the Gods of war we went when we did! ♦

Without the work of the forecasters allowing the invasion to go ahead at the right time, the outcome of the war might have been very different.

The Great Fire of London

You might not immediately think about the wind, but the weather played a pivotal role in the Great Fire of London, which began on 2 September 1666 and lasted for five days, destroying about one-third of the city and making approximately 100,000 people homeless. It had been a long, hot summer, which meant the area around Pudding Lane where the fire started was tinder-dry and highly flammable. Perhaps more significantly, however, was the strong easterly wind that developed at the beginning of the fire, causing it to spread easily from house to house in the narrow, tightly packed streets. It wasn't until the winds dropped on the third night that the amateur firefighters were able to get the fire under control, finally extinguishing it fully on the fifth day.

Kites and the Weather

Kites have been used by humans as tools for millennia, with the earliest depiction of a kite dating back to cave art in Indonesia from around about 9500 to 9000 BCE. More recently, in the nineteenth century, Benjamin Franklin famously used a kite in his experiments to prove that lightning was a form of electricity, and the Wright Brothers and other early aeronautical pioneers used them in their research into flight.

From a meteorological point of view, upper-air observations of the atmosphere have been of interest since the first thermometers were flown on kites in the mid-seventeenth century, and they were still used for routine observations into the early twentieth century. In particular, their use was revived by English scientists William Henry Dines and Sir Napier Shaw, who conducted a series of experiments with meteorological kites on behalf of the Met Office in the early 1900s at Pyrton Hill in Oxfordshire. They were very advanced in their techniques, and Dines published several important papers outlining his findings that contributed a great deal to advancing our knowledge of the weather. But even though the experiments were an immediate success, using kites had obvious limitations, and they were soon replaced by weather balloons, which were generally acknowledged to be more effective.

Optimum wind speed is about 8 to 24mph

The Perfect Kite-flying Conditions

Kite flying is one of best things to do on a windy day, at least according to Mary Poppins, although many of us have probably found ourselves frustrated by our attempts to keep our kites aloft. So, what are the best kite-flying conditions? The key ingredient is of course wind, but too much or too little will scupper your efforts. The optimum wind speed is about 8 to 24 miles per hour, depending on the type of kite you are flying. A good general rule of thumb is to test whether you can feel the breeze on your face. The direction of the wind is also important. You can use trees, flags or other everyday objects that are affected by the wind to gauge its direction, or you can throw some blades of grass in the air and watch where they land. Once you have found the direction of the wind, you should stand with your back against it. If you can feel wind equally on the backs of both ears, you are in the perfect position.

The Perfect Clothes-drying Weather

Drying our clothes outside on a washing line is both economical and efficient, but only if the weather conditions are right. There's obviously no point hanging them out in the rain, but is sunny or windy weather better? Although the heat from the sun can help to evaporate the water from our clothes, wind is also important. The wind prevents the air from becoming saturated in still, humid conditions, which can slow the drying of your clothes if the air is unable to hold more water vapour. The ideal, therefore, is a cold, dry, windy day rather than a hot and humid one – in other words, it's easier to dry your clothes outside in the UK than it is in the tropics.

foggy [faw-gee]

adjective thick with or having much fog; misty

origins: 1300–50; Middle English *fogge, fog*; compare Norwegian *fogg* 'long grass on damp ground'

FOGGY

Fog is cloud formed at ground level that causes a reduction in visibility to less than 1,000 metres. It becomes a much more noticeable and thick fog when visibility drops below 180 metres, while severe disruption to transport occurs when the visibility falls below 50 metres over a wide area, referred to as dense fog.

Also Known As…

PEA-SOUPER

HAAR

FRET

MIZZLE

SMOG

MIST

HAZE

What Causes Fog?

Fog is caused by tiny water droplets suspended in the air. When air is cooled close to the Earth's surface, water vapour condenses onto small particles known as condensation nuclei. These droplets are light enough to remain suspended in the air and create the cloud-like fog. The thickest fogs tend to occur in industrial areas, where there are more pollution particles in the air, allowing water droplets to coalesce and grow.

What Is the Difference Between Fog, Mist and Haze?

Fog, mist and haze all affect visibility, which can have an impact on many aspects of life, from driving conditions to shipping and aviation. Fog and mist differ by how far you can see through them. In our meteorological glossary, fog is defined as 'obscurity in the surface layers of the atmosphere, which is caused by a suspension of water droplets'. By international agreement (particularly for aviation purposes), fog is the name given to resulting visibility of less than one kilometre. However, in forecasts for the public, fog generally refers to conditions when visibility is less than 200 metres.

Mist, meanwhile, is defined as 'when there is such obscurity and the associated visibility is equal to or exceeds 1,000 metres'. Like fog, mist is still the result of the suspension of water droplets but simply at a lower density. Mist is typically quicker to dissipate and can rapidly disappear with even slight winds. It's also what is visible when you can see your breath on a cold day.

Haze, on the other hand, is caused by a different process to mist and fog. Instead of the suspension of water droplets in the air, haze is the suspension of extremely small, dry particles, such as dust or smoke. These particles are invisible to the naked eye, but in sufficient quantity they can give the air an opalescent appearance. They can also contribute to creating a red sky at sunrise or sunset.

Why Is it Often Foggy at the Coast?

Coastal fog is usually a result of advection fog, which forms when relatively warm, moist air passes over a cool surface. In the UK, the most common occurrence of coastal fog is when warm air moves over the cool surface of the North Sea towards the east coast of the UK, where it is known as 'haar' in Scotland and 'fret' in some eastern parts of England. When this happens, the cold air just above the sea's surface cools the warm air above it until it can no longer hold its moisture. This forces the warm air to condense, forming tiny particles of water that create the fog that we see.

Coastal fog usually occurs in the spring and summer months when conditions begin to warm up but the sea (which warms more slowly) stays relatively cold. The impact, location and movement of coastal fog depends on a number of conditions, including wind strength, wind direction and land temperature. If, as is common along the UK's east coast, the winds blow in from the east, the coast will be rapidly covered in a blanket of fog. If the land temperature is warm, the fog can quickly dissipate as the parcel of air warms. However, if the land temperature is cooler, the fog can linger for a longer time. Coastal fog can also refer to pre-existing fog that is transferred from a distant source and is simply moved to the coast by prevailing weather patterns.

Why Do People Drive Faster in Fog?

When driving, most people are aware of the need to slow down in foggy conditions because of how dangerous it can be, but did you know fog has a hazardous impact on your speed perception? One way our brains judge speed is by contrasting our movement with our surroundings, such as trees or buildings flashing past in our peripheral vision. But in foggy conditions the contrast is greatly reduced, giving the impression you are driving slower than you actually are. Many drivers increase their speed as a result.

Five Types of Fog

Although it might be difficult to tell the difference if you are caught in fog, there are in fact five distinct types. You'll never see fog in the same way again.

1 **RADIATION FOG** usually occurs in the winter, aided by clear skies and calm conditions. The cooling of land overnight by thermal radiation causes the temperature of the air close to the surface to drop, which reduces the ability of the air to hold moisture. Condensation and fog occurs that usually dissipates soon after sunrise as the ground warms – except in high-elevation areas where the sun has less influence in heating the surface.

2 **VALLEY FOG** forms where cold, dense air settles and condenses in the lower parts of a valley. It is the result of temperature inversion, with warmer air passing above. Valley fog is confined by local topography and can last for several days in calm conditions during the winter.

3 **ADVECTION FOG** occurs when moist, warm air passes over a colder surface and is cooled. Common examples are when a warm front passes over an area with snow cover, or at sea when moist tropical air moves over cooler waters. If wind blows in the right direction, sea fog can be transported over coastal land areas.

4 **UPSLOPE FOG (OR HILL FOG)** forms when winds blow air up a slope in a process called orographic uplift. The air cools as it rises, allowing the moisture in it to condense.

5 **EVAPORATION FOG** is caused by cold air passing over warmer water or moist land. It often causes freezing fog, or sometimes frost. When the relatively warm water evaporates into low air layers, it warms the air, causing it to rise and mix with the cooler air that has passed over the surface. Warm, moist air cools as it mixes with colder air, allowing condensation and fog to occur. Evaporation fog can be one of the most localised forms of fog. It can happen when cold air moves over heated outdoor swimming pools or hot tubs, where steam fog easily forms, or when cold fronts or cool air masses move over warm seas – this often occurs in autumn when sea temperatures are still relatively warm after the summer but the air is already cooling.

1

Ground radiates heat

Air close to the ground cools and condenses into fog

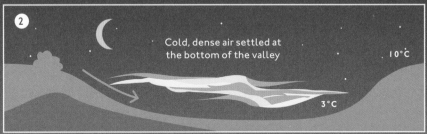

2

Cold, dense air settled at the bottom of the valley

10°C

3°C

3

Fog forms

Moist warm air moves over cold surface

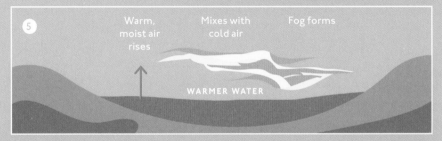

4

Fog forms on the slope

Humid air is forced to rise up when it hits a slope

The air cools and condenses

SLOPE

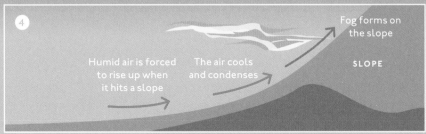

5

Warm, moist air rises

Mixes with cold air

Fog forms

WARMER WATER

What is a Fog Bow?

A fog bow is a rainbow that appears in fog rather than during rainfall. Fog bows are fainter and slightly harder to see than rainbows, but the complete circle of the optical effect is more likely to be visible. Because the water droplets that make fog are smaller than raindrops, the full spectrum of colours in a fog bow are hard to distinguish and generally appear as a bow of faint red and blue.

Why Is It Foggy on Bonfire Night?

You may have noticed that it is often foggy. This isn't anything to do with the weather on that day always being the same each year. Because of all of the fireworks and bonfires, there are lots of smoke particles in the atmosphere, which act as condensation nuclei for water molecules to attach themselves to, bringing a greater chance of fog.

How Much Water Is in Fog?

Fog is made up of tiny water droplets all suspended in what is essentially a cloud on the ground, containing up to 0.5 millilitres of water per cubic metre. If you were to fill an Olympic-sized swimming pool with fog and then somehow condense it, you would be left with around 1.25 litres of water, which wouldn't be enough to fill an average kettle.

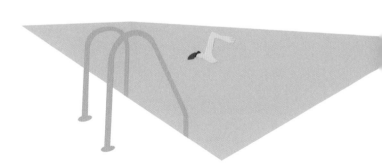

How Do We Measure Visibility?

Today, visibility sensors (*pictured below*) measure the meteorological optical range, which is defined as the length of atmosphere through which a beam of light travels before its power is reduced to 5 per cent of its original value. However, for a great many years, meteorological visibility was estimated by a human observer judging the appearance of distant objects against a contrasting background, usually the sky. Various rules were established for how visibility should be judged:

- An object should not merely be seen but should be identifiable against the background as a specific object.

- Visibility should be estimated at ground level, where there is an uninterrupted view of the horizon.

- If the visibility varies from one direction to another, the lowest value should be reported.

Fog Warnings

Fog can be so severe that it is subject to Met Office National Severe Weather Warnings. It is especially dangerous to transport networks, as it can dramatically reduce visibility on the roads, and there have been numerous examples of large motoring accidents due to fog in the UK over the years. It is such a hazard that it became mandatory for all cars made after 1 October 1979 to have rear fog lights fitted as standard. But don't forget to turn them off again after the fog has cleared – it's illegal to use fog lights in normal driving conditions.

Why Is Fog So Difficult to Forecast?

The formation of fog is dependent on the time of year, how much cloud cover there is, the strength of the wind, temperature, the amount of moisture in the air and geographical location. With so many factors in play, it can be a tricky thing to forecast.

In the case of radiation fog, the ideal conditions for its formation are clear skies, light winds and plenty of moisture. If there is a high moisture content in the air, the temperature will not have to fall as far to reach the fog point and there will be a greater risk of fog forming. This is often why fog forms more readily around lakes, rivers and reservoirs. In addition, winds need to be gentle enough to keep the moisture in the air at the right concentration. Too light and it'll form dew; too strong and the moisture will be mixed through the air and probably keep temperatures too high. And clear skies at night allow temperatures to drop as heat escapes more readily from the Earth's surface.

With the right wind, moisture and lack of cloud, the fog point will be reached. But we must also consider the amount of sunshine and how strong the sun is at a particular time of year; whether there will be any rain; how much heat the ground has absorbed; and so on. If any of those variables changes, even by a small amount, it will have a knock-on effect on our ability to predict fog, highlighting why it is one of the hardest weather phenomena to forecast.

Ingredients for a Long-lasting Fog

° Lots of moisture in the air, perhaps close to a water source

° Low overnight temperatures

° Little cloud cover

° Light winds

° Weak sunshine to prevent the fog from being 'burnt off'.

Fog + Smoke = Smog

Britain has long been affected by mists and fogs, but these became much more severe after the onset of the Industrial Revolution in the late 1700s. Factories pumped gases and huge numbers of particles into the atmosphere, which in themselves could be poisonous. The pollutants in the air, however, could also act as catalysts for fog, as water can cling to the tiny particles to create polluted fog – and smoke plus fog is known as smog. When some of the chemicals mix with water and air, they can turn into acid, which can cause skin irritations, breathing problems and even corrode buildings. Smog can be identified easily by its thick, foul-smelling, yellow, green or brown characteristics, totally different to the clean white fog in country areas – 'pea-souper', a colloquial term for smog, captures some of this noxious quality.

One of the earliest reports of a polluted fog comes from London, which was blanketed with a thick smog, smelling of coal tar, in December 1813. Lasting for several days, people claimed you could not see from one side of the street to the other. A similar fog in December 1873 saw the death rate across London rise by 40 per cent. And marked increases in death rate occurred after the notable smogs of January 1880, February 1882, December 1891, December 1892 and November 1948. The worst-affected area of London was usually the East End, where the density of factories and homes was greater than almost anywhere else in the capital. The area was also low-lying, making it hard for any smog to disperse.

Other British cities were also badly affected by smog. Edinburgh was known as 'Auld Reekie', or 'Old Smoky', because of the polluted fog that blanketed the city as a result of the smoke pouring from the factories and open fires in the tenements of the overcrowded Old Town. And the highest volume of coal was burned in the industrial heartlands of the Midlands, the north-west of England and the south-west of Wales, so although no records of air pollution were kept during the nineteenth and early parts of the twentieth centuries, these areas were most likely to be adversely affected by smog.

The Great Smog of 1952

By the middle of the twentieth century, smog had become a common part of London life, but nothing quite compared to the smoke-laden fog that shrouded the capital from Friday, 5 December to Tuesday, 9 December 1952. During the day on 5 December, the fog was not especially dense to begin with and generally possessed a dry, smoky character. When nightfall came, however, the fog thickened, and visibility dropped to only a few metres. The following day, the sun was too low in the sky to evaporate the fog away. That night and on the Sunday and Monday nights, the fog again thickened. In many parts of London, it was impossible for pedestrians to find their way at night, even in familiar districts. In the Isle of Dogs area, the fog was so thick that it was reported people could not see their feet.

The weather in November and early December 1952 had been very cold, with heavy snowfalls across the region. To keep warm, the people of London were burning large quantities of coal, and smoke was pouring from the chimneys of their houses. Under normal conditions, this smoke would have risen into the atmosphere and dispersed, but an anticyclone was hanging over the region. This pushed the air downwards, warming it as it descended. This created an inversion, where air close to the ground was cooler than the air higher above it. So, when the warm smoke rose from London's chimneys, it was trapped. The inversion of 1952 also trapped particles and gases emitted from factories, along with pollution that winds from the east had brought from industrial areas on the Continent. During the period of the smog, huge amounts of impurities were released into the atmosphere. On each day during the foggy period, it is estimated that the following pollutants were emitted: 1,000 tonnes of smoke particles, 2,000 tonnes of carbon dioxide, 140 tonnes of hydrochloric acid and 14 tonnes of fluorine compounds. In addition, and perhaps most dangerously, 370 tonnes of sulphur dioxide were converted into 800 tonnes of sulphuric acid.

The fog finally cleared on 9 December, but it had taken a heavy toll:

- About 4,000 people were said to have died as a result of the fog, but it could have been many more, as mortality rates were elevated for months after the smog had cleared.

- Many people suffered from breathing problems.

- Press reports claimed cattle at Smithfield meat market had been asphyxiated by the smog.

- Travel was disrupted for days.

CLEANING UP OUR ACT

The Great Smog of 1952 was so severe that it led to legislative change, with a series of laws enacted to avoid something similar happening again in the future. These included the Clean Air Acts of 1956 and 1968, which banned emissions of black smoke and decreed residents of urban areas and operators of factories must convert to smokeless fuels.

People were given time to adapt to the new rules, however, and fogs continued to be toxic for some time after the Act of 1956 was passed. In 1962, for example, 750 Londoners died as a result of a polluted fog. Although poor air quality is still a problem in many British cities, nothing on the scale of the deadly 1952 Great Smog has ever occurred again, thanks partly to this and subsequent pollution legislation and also to modern developments, such as the widespread use of central heating.

London Fog and Claude Monet

Although for many people the London smog was a blight on the capital, it also proved to be inspirational for a number of French artists, including Claude Monet, dubbed the 'Father of Impressionism'. Monet made multiple trips to London during the course of his life, first when fleeing the Franco-Prussian War and later when he and Impressionism were fully established. He painted almost 100 views of the Thames, many from his rooms on the fifth and sixth floors of The Savoy Hotel, where he could look eastwards to Waterloo Bridge and the South Bank and west towards Charing Cross Bridge. The fog and pollution gave the air and light a semi-opaque quality that lent itself to Monet's impressionistic style, and he famously said:

Without fog London would not be beautiful.

pouring [pohr-ing]

adjective heavy and continuous rainfall

origins: 1300–50; Middle English *pouren*; root uncertain

POURING

It's raining. It's pouring.
The old man is snoring.
He went to bed and bumped his head
and didn't get up in the morning.

We are no strangers to rain in the UK, and if we didn't get out of bed every time it was pouring, we wouldn't get very much done. In fact, it is probably the most influential element of the British weather in terms of the day-to-day decisions we make – from the clothes we wear to whether or not we hang out the washing.

What is Rain?

At its simplest, rain is drops of liquid water falling from the sky. Clouds form when water vapour (a gas) cools and condenses to form water droplets (a liquid) but these droplets are so tiny, they stay in the sky. To become rain, the water droplets must grow by acquiring more water and becoming larger. Some droplets collide with others in order to become larger, while others grow as water condenses out of the air into the droplet. When these drops become too heavy to stay in the cloud, we get rain. The size of raindrops is highly variable, from as small as 0.5 millimetres in diameter to 6 millimetres.

Also Known As…

Perhaps not surprisingly, considering how accustomed to wet weather we are in the UK, there is no shortage of British words and phrases to describe the rain.

SPITTING

PICKING (Welsh for spitting)

POURING

HEAVING/ STOATING/ CHUCKING/ TIPPING/ BUCKETING (IT) DOWN

COMING DOWN IN STAIR RODS

THAT FINE RAIN

TEEMING

SMIRR (Scots for a light, mist-like rain)

NICE WEATHER FOR DUCKS

THE HEAVENS HAVE OPENED

RAINING SIDEWAYS

A DELUGE

ACHE AND PAIN (Cockney rhyming slang)

FISS (Scottish for drizzle)

DROOKIT (Scottish for extremely wet)

LETTY WEATHER

PLOTHERING

How Much Does It Rain in the UK?

The wettest parts of the UK are concentrated in mountainous regions, with observation sites in Snowdonia, the Lake District and the Scottish Highlands all receiving more than 4 metres of rainfall in a year.

- North-west England, especially the Lake District in Cumbria and western-facing slopes of the Pennines.

- Western and Mid-Wales, particularly the mountainous Snowdonia region in the north.

- South-west England, mainly the higher elevation areas of Dartmoor, Exmoor and Bodmin Moor.

- Parts of Northern Ireland, mainly higher elevation areas such as the Sperrin Mountains, Antrim Mountains and Mourne Mountains.

These areas all have common characteristics, given their high elevations (or even mountainous status) and their northern or western locations in the UK.

RAINFALL AMOUNT ANNUAL AVERAGE 1981–2010

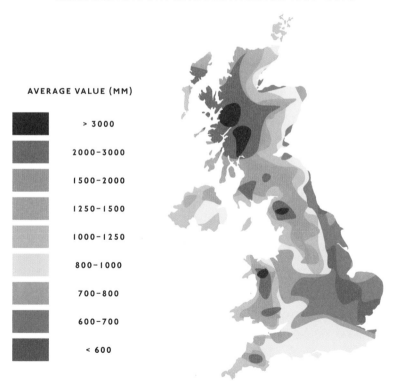

AVERAGE VALUE (MM)

	> 3000
	2000–3000
	1500–2000
	1250–1500
	1000–1250
	800–1000
	700–800
	600–700
	< 600

POURING

WINTER
2013–14

MONTH
October 1903

SPRING
1947

YEAR
1872

Britain's
Wettest

SUMMER
1879

DAY AT WIMBLEDON
62.7mm on
28 June 1906

DAY
341.4mm of rainfall
on 5 December 2015
at Honister Pass
in Cumbria

AUTUMN
2000

Britain's Biggest Soaking

The wettest day recorded in Britain was 5 December 2015 when Honister Pass in Cumbria recorded 341.4 millimetres of rain in a single day, the result of Storm Desmond, which brought widespread heavy rain and storm-force winds to areas of Scotland and northern England. The same rainfall event also saw the highest accumulations ever recorded in Britain for a two-day period (405 millimetres), recorded nearby in Thirlmere.

How Much Rain Falls Every Year in the UK?

The volume of rain that falls across Scotland in an average year is enough to fill Loch Ness, the largest lake by volume in the UK, more than 17 times over:

AREA OF SCOTLAND: 82,000 km²
DEPTH OF ANNUAL RAINFALL: 0.0016 km
APPROX. VOLUME OF LOCH NESS: 7.4 km³
(82,000 x 0.0016) / 7.4 = 17.7

The volume of rain that falls across Northern Ireland in an average year is enough to fill Lough Neagh, the largest lake by area in the UK, around four and a half times:

AREA OF NORTHERN IRELAND: 14,130 km²
DEPTH OF ANNUAL RAINFALL: 0.00113296 km
APPROX. VOLUME OF LOUGH NEAGH: 3.528 km³
(14,130 x 0.00113296) / 3.528 = 4.5

And the volume of rain that falls across England on an average day is enough to fill the Royal Albert Hall almost 3,000 times:

AREA OF ENGLAND: 130,000 km²
DEPTH OF DAILY RAINFALL: 0.0000023 km
APPROX. VOLUME OF THE ROYAL ALBERT HALL: 0.0001 km³
(130,000 x 0.0000023) / 0.0001 = 2,990

While the volume of rain that falls in Wales on an average day is enough to fill the bowl of the Millennium Stadium more than 55 times.

AREA OF WALES: 20,735 km²
DEPTH OF DAILY RAINFALL: 0.00000399704109 km
APPROX. VOLUME OF THE MILLENNIUM STADIUM: 0.0015 km³
(20,735 x 0.00000399704109) / 0.0015 = 55.3

POURING

Different Types of Rain

Rain is classified according to how it is generated, with three main types of rainfall:

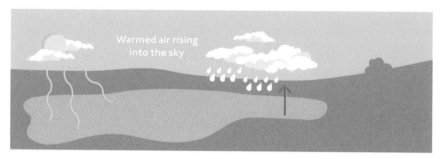

CONVECTIVE RAIN

This is rain that is caused by rising pockets of warm air. When this air cools the further up into the atmosphere it travels, it becomes unstable, and the vapour in the air condenses into shower clouds (*see* 'Showery').

FRONTAL RAIN

This occurs when warm air meets cold air in a weather front. The warm air rises above the cold air and begins to cool. As a result, the water vapour condenses into water, forms a cloud and eventually falls as raindrops. Frontal rain can happen anywhere in the UK, and it usually brings grey overcast skies and persistent rain through a large portion of the day.

OROGRAPHIC RAIN

When the wind pushes air towards a hill, it is forced upwards. When it reaches the top of the hill it cools down (again, the air higher up in the atmosphere is colder), and if it has enough moisture, the water vapour will condense into liquid water to form a cloud and eventually fall as raindrops. This is why we tend to see more rain on the hills than elsewhere.

Why Are Some Places Wetter Than Others?

There is a clear divide between how much it rains in the north-west and south-east of the UK. The prevailing warm, moist westerly winds mean that the west of the UK is more likely to receive rainfall from Atlantic weather systems, in the form of frontal rainfall. These weather systems usually move from west to east across the UK and as they do so the amount of rainfall they deposit reduces.

The mountains of northern and western UK also force the prevailing westerly winds to rise, which cools the air and consequently enhances the formation of cloud and rain in these locations (this is known as orographic enhancement). Of course, frontal and orographic rainfall are not the only rainfall mechanisms, but they are the most common in the UK.

BRITAIN'S WETTEST AREA

Argyllshire in Western Scotland, with 2274.9mm on average per year

What Is Rain Shadow?

When it comes to average annual rainfall, there is huge variation from west to east and from high to low across the UK, with some upland areas in western Britain soaking up more than three times the average rainfall each year compared to lower-lying eastern areas. That's because once the prevailing westerly wind uses up most of its moisture over the hills in the west, it's much drier when it arrives in the east. When the weather is normally drier downwind of hills or mountains, we call it a rain shadow.

Wet Weather Record Holders

Cumbria is officially crowned the chucking-it-down champion.

UK RAINFALL RECORDS FOR CONSECUTIVE
RAINFALL DAYS (0900–0900 GMT)

DAYS	RAINFALL (MM)	DATE	LOCATION
Highest 2-day total	405.0	4 to 5 Dec 2015	Thirlmere (Cumbria)
Highest 3-day total	456.4	17 to 19 Nov 2009	Seathwaite (Cumbria)
Highest 4-day total	495.0	16 to 19 Nov 2009	Seathwaite (Cumbria)
Highest monthly total	1396.4	1 to 31 Dec 2015	Crib Goch (Snowdon)

AREA	TOTAL RAINFALL ON AVERAGE IN 1 YEAR (MM)
Argyllshire	2274.9
Dunbartonshire	2066.5
Inverness	2034.8
Merionethshire	1914.0
Ross and Cromarty	1858.1
Carnarvonshire	1809.7
Buteshire	1771.2
Kirkcudbrightshire	1696.7
Westmorland	1652.6
Brecknockshire	1643.7

POURING

How Big Is a Raindrop?

TYPICAL CLOUD DROPLET:
0.01mm

TYPICAL DROPLET OF DRIZZLE:
0.1mm

**NUMBER OF CLOUD DROPLETS
IN A DROPLET OF DRIZZLE:**
10

**NUMBER
OF CLOUD
DROPLETS IN
A RAINDROP:**
400

**TYPICAL
RAINDROP:**
4mm

**NUMBER
OF CLOUD
DROPLETS IN
THE LARGEST
RECORDED
RAINDROP:**
880

**LARGEST RECORDED
RAINDROP:**
8.8mm

What Is Drizzle?

Drizzle is a type of liquid precipitation consisting of very small droplets of water falling from low-level stratus clouds. To be classed as drizzle, the droplets must be less than 0.5 millimetres in diameter. It is obviously difficult to measure a water droplet in everyday life, but a useful rule of thumb is that raindrops will make a splash when landing in a puddle, whereas drops of drizzle won't. The droplets in drizzle are larger than the droplets in the clouds, but smaller than raindrops. For the drizzle to form, the cloud must be fairly low to the ground – usually below 460 metres from the surface.

Why Does It Drizzle?

In the UK, we're very familiar with drizzle. Our prevailing winds from the Atlantic pick up lots of moisture and therefore clouds as they travel over the sea. You need low-level stratus or stratocumulus clouds to experience drizzle on the ground, because drizzle falling from clouds that are at medium or high levels in the atmosphere will almost always evaporate before reaching us due to the small size of the droplets. Convective clouds, such as cumulus or cumulonimbus, do not produce drizzle because the updrafts in them are more vigorous and keep the small droplets of drizzle within the cloud until they form larger raindrops. Thin stratus or stratocumulus have weaker updrafts, releasing the smaller droplets of water sooner.

There are even weaker updrafts over the sea and ocean than over land, which is why drizzle is more common over the sea or near to the coast. And low cloud is more common the higher up a hill or mountain you go, so drizzle will also occur more frequently on higher ground. This is because air that is close to saturation can be forced to rise by higher ground, resulting in condensation and low cloud and drizzle.

Rain As a Decoy

When radar was first used during the Second World War, precipitation was the noise masking the signal from enemy targets. Radio waves emitted from a radar antenna travelled outwards at the speed of light, only bouncing back to the same antenna when they encountered an object. When the weather was clear, radar could quickly provide intelligence on the position of aircraft in the sky. When it was raining or snowing, echoes visible from the radar signal would only tell us about the weather — allowing the enemy to use bad weather as a cover for their operations.

The Smell of Rain

Some scientists think that our reliance on rain in cultures throughout history may be the reason why so many people enjoy the smell of rain — yes, believe it or not, rain has a smell. When raindrops fall on dusty or clay soils, they trap tiny air bubbles on the surface which then shoot upwards — as in a glass of champagne — and burst out of the drop, throwing aerosols of scent into the air, where they are then distributed by the wind. This is what is responsible for the familiar smell of rain, and it is called petrichor. The word comes from the Greek words *petra*, meaning stone, and *ichor*, which in Greek mythology refers to the golden fluid that was said to flow through the veins of the gods and the immortals.

Petrichor was first defined by two researchers at the Australian Commonwealth Scientific and Industrial Research Organisation in a 1964 article for the journal *Nature*. In their research, rocks and soil that had been exposed to warm, dry conditions were shown to contain a yellow-coloured oil that had become trapped. The source of this oil is a combination of oils secreted by plants during dry weather and chemicals released by soil-dwelling bacteria. When a higher humidity is experienced as a precursor to rain, the pores of rocks and soil become trapped with moisture, forcing some of the oils to be released into the air. But the strongest smell is released when rainfall arrives. The release of the scent is most prominent when light-to-moderate rain falls on sandy or clay soils. During heavy rain, the speed of the drops represses the creation of bubbles, stopping the release of the aerosols.

How to Draw a Raindrop

While raindrops are usually represented in the shape of a teardrop, in reality they are not (you'll have to excuse the raindrop shapes on the cover of this book). When they first form high up in the atmosphere, they are a spherical shape, as the water molecules bind together held by surface tension. As they begin to fall, their shape changes as they hit other raindrops, while air resistance causes the bottom of the drop to flatten and curve, resembling the shape of a jelly bean. Perhaps 'it's raining jelly beans' should replace the classic 'cats and dogs'.

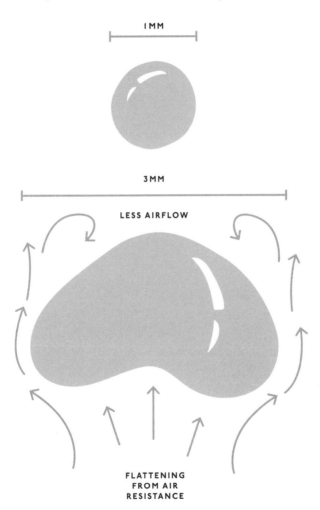

How Is Rain Forecast?

Along with our understanding of weather fronts and convection, advancements in satellite imagery and in particular radar have vastly improved how we forecast rain. Following an upgrade to the Met Office's thirty-year-old rainfall radar network in 2018, we can now gain more real-time information from falling precipitation – rain, sleet, snow, hail and ice pellets – than ever before, including the size and shape of the droplets. The UK is now covered by fifteen radar sites, collecting data at one-kilometre resolution every five minutes. However, radar has not been without its problems. Anything from birds and insects in the sky to hills and buildings on the ground has interfered with the vertical radar signal and returned false data.

Now, taking advantage of the UK's upgraded radar network, we can send two radar beams at once, using a technology known as dual-polarisation. One beam travels horizontally and another travels vertically through the atmosphere so we can visualise objects in three dimensions. By picturing the size and shape of objects in the sky, we can discount flying creatures and ground clutter. We can also draw accurately shaped precipitation particles. Our radar images might not be works of art, but they can identify rain, snow, hail and ice pellets as well as distinguish small raindrops from large drops. It's crucial to be able to tell the difference when extreme weather hits.

Measuring Rainfall

The first known rainfall records were kept by the Ancient Greeks in 500 BC and in India at a similar time period. With rainfall in abundance in the British Isles, it wasn't seen as such a valuable thing, at least until the Enlightenment when scientists wanted to quantify and measure all sorts of elements.

Many different types of rain gauge have been designed and used over the years, most consisting of a circular collector and a funnel that channels the collected rain into a measuring mechanism. The entrance to the gauge through the funnel tends to be narrow to avoid debris clogging the mechanism and undesirable evaporation in hot weather.

STORAGE RAIN GAUGE

Since the earliest years of weather records, the de facto standard for the measurement of daily rainfall has been the 0900 GMT reading made by an observer from a storage rain gauge. The gauge has a sharp brass or steel rim 12.7 centimetres in diameter, sited 30 centimetres above ground level with a funnel that collects rain in a narrow-necked bottle placed in a removable can. To make the rainfall measurement, the observer empties the collected rain into a graduated glass rain measure. Versions of the 12.7-centimetre gauge with greater storage capacity are used at more remote sites, where readings are taken less often.

TIPPING-BUCKET RAIN GAUGE

For many years the Met Office has used a tipping-bucket rain gauge for the automatic measurement of rainfall rate. The collecting funnel has a sampling area of 750 centimetres squared, the rim is set 450 millimetres above the surrounding ground level and a mechanism records each time a rainfall increment of 0.2 millimetres has been detected.

How to Make Your Own Rain Gauge

If you want to try your hand at becoming a weather observer, a rain gauge is a good family-friendly way to start.

YOU WILL NEED

An empty plastic
    bottle (a two-litre
    fizzy drinks bottle
    would be ideal)

Scissors

Jelly (three or four
    cubes made up
    as directed on
    the packet)

Sticky tape

Ruler

Paper

Pencil

WHAT TO DO

Cut around the plastic bottle about two-thirds of the way up.

Your bottle needs a flat bottom to be able to measure the rainfall properly. Pour a few centimetres of jelly into the bottle to create a flat bottom. Let the jelly set.

Turn the top part of the bottle upside down and place it inside the bottom part to act as a funnel – fix it in place using the tape.

Make a scale in centimetres on another piece of tape, using a ruler, and fix it to the side of your bottle. Use the top level of the jelly as your starting point for the piece of measured tape.

Find a place outside to put your rain gauge. It must be in the open and away from trees.

Dig a hole and bury your rain gauge so that the top is sticking out of the ground. This will stop it from blowing over on windy days.

Check the rain gauge every day at the same time, measure the amount of rain collected and empty the bottle.

Write down the amount of rain collected in your weather diary every day.

How Fast Does a Raindrop Travel?

The speed of a raindrop is largely determined by its size, with smaller drops falling more slowly than larger drops. It is therefore difficult to give an exact figure as to how long it takes a raindrop to reach the ground, as the height at which raindrops fall and their size vary widely. However, if we say that raindrops fall at an average speed of around 14 miles per hour, and assuming a cloud base height of around 760 metres, a raindrop would take just over two minutes to reach the ground. Larger raindrops can fall as fast as 20 miles per hour, while the smallest raindrops can take up to 7 minutes to reach the earth.

How Accurate Is Weather Lore About the Rain?

COWS LIE DOWN WHEN IT'S GOING TO RAIN

A Met Office survey revealed that 61 per cent of people in the UK believe that cows lying down in a field is a sign that rain is on the way. Several theories have been proposed for this: some people say that cows are particularly sensitive to atmospheric pressure, while others have suggested that they sense the moisture in the air and lie down to save themselves a dry patch of grass. But the truth is that cows lie down for many reasons, and there's no scientific evidence that rain is one of them.

RAIN BEFORE SEVEN, FINE BY ELEVEN

Weather systems tend to be variable and move through the UK fairly quickly, as a result of the prevailing westerly airflow off the Atlantic. While this can sometimes mean that a weather front passes over the UK during the course of a morning, this is not always the case and rain can (and often does) stay around for longer than a morning. So, on the occasions when this saying is accurate, it is probably just a coincidence, and it is not one to rely on.

ST SWITHUN'S DAY

The saying goes:

St Swithun's Day, if thou dost rain,
For forty days it will remain;
St Swithun's Day, if thou be fair,
For forty days 'twill rain no more

This proverb originates with St Swithun, a bishop of Winchester in Anglo-Saxon times, who died in 863. Before his death, he requested that he be buried outside, where he said he might be subject 'to the feet of passersby and to the raindrops pouring from on high'. More than a century later, his body was moved to an indoor shrine, at which point a heavy shower began, supposedly as a result of the saint's anger at being moved. The rain was said to have lasted for 40 days and nights. Although there is no record of when the move actually took place, folklore has attributed it to 15 July, St Swithun's Day, leading to the proverb that rain on 15 July will continue for 40 days and nights.

The location of the jet stream shortly after the summer solstice does in fact influence the following summer's weather, which may be one of the reasons why the saying took hold. If the jet stream is located in a southerly position, it is likely to be a more unsettled summer. If the jet stream is in a northerly position, the weather is likely to be brighter and dry throughout summer. In fact, there has been no occurrence of rainfall for such a prolonged period of time since records began, and the proverb is more fiction than fact.

The lesson from this is that you should probably check the Met Office forecast rather than relying on folklore!

POURING

stormy [stawr-mee]

adjective affected, characterised by or subject to storms; tempestuous

origins: 1150–1200; Middle English; Old English *stormig*

STORMY

Also Known As...

BARBER: a storm at sea, carrying sleet, snow or spray when the temperature is below zero, known for freezing the decks of boats

BLUNK (England): a sudden squall

DOISTER (Scotland): a severe storm from the sea

GOWK STORM (Scotland and Ireland): a brief storm occurring towards the end of April or beginning of May

PEESWEEP STORM (England and Scotland): an early spring storm

The British Isles are no strangers to storms, otherwise known as extratropical cyclones, depressions or deep lows, areas of rotating winds spiralling towards a low-pressure centre.

While warmer climes are affected by tropical cyclones – including hurricanes and typhoons – the UK is only affected by extratropical cyclones. These are cyclones that occur outside the tropics and subtropics, typically in the midlatitudes around 30 to 60 degrees north or south of the equator. For this reason, they are also sometimes called midlatitude cyclones.

Under Pressure

Atmospheric pressure is the force exerted by the weight of air above us, and it is measured in hectoPascals (hPa), which are also sometimes called millibars. Standard pressure at sea level is defined as being 1013 hPa on average, but large areas with either higher or lower pressure than this can develop. These areas are all relative to one another, so what defines a high will change depending on the area around it. On a weather chart, lines joining places with equal sea-level pressures are called isobars. Charts showing isobars are useful because they identify features such as anticyclones (areas of high pressure) and depressions (areas of low pressure).

Areas of high and low pressure are caused by ascending and descending air. As air cools, it falls, leading to high pressure at the surface and the suppression of weather development, often resulting in calm, clear or sunny conditions. As air warms, it rises, leading to low pressure at the surface and causing winds to circulate rapidly inwards and upwards. As the air ascends further, it cools, often leading to the formation of clouds and precipitation. Storms are caused by deep areas of low pressure that bring strong winds and heavy rain.

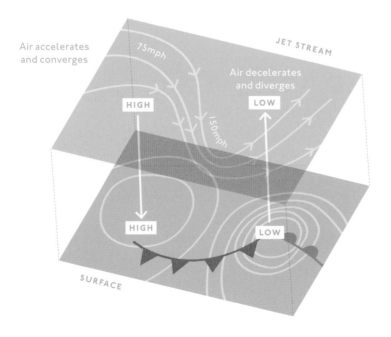

How Do Storms Form?

Situated on the edge of the Atlantic, the British Isles are sometimes subject to low-pressure systems that can bring wet, windy and stormy weather to our shores, particularly during the winter months. While tropical cyclones form in regions of relatively uniform temperature, midlatitude depressions are formed at the boundary between warm air and cold air in the midlatitudes. Significant changes in wind speed and direction can also be seen at different heights in the atmosphere when a midlatitude depression forms, a phenomenon known as wind shear.

The strongest wind speeds tend to be found much higher in the atmosphere at the jet streams, fast-flowing currents of air at around 8 to 11 kilometres above sea level. The winds of a jet stream are often defined as exceeding 69 miles per hour (60 knots) and can be as fast as 249 miles per hour. There are five main jet streams around the world: the subtropical jet stream and the polar jet stream in each hemisphere as well as the equatorial easterly jet over the Indian Ocean and the African continent. The polar jet streams are caused by significant temperature contrasts – where warm air meets cold air. In some areas close to a maximum of the jet stream – known as jet streaks – areas of acceleration and deceleration cause air from below to rise. This lowers atmospheric pressure below and the air starts to rotate, forming a cyclone. This is known as cyclogenesis.

Midlatitude depressions have cold cores. They derive their energy from horizontal temperature differences in the atmosphere – the boundary between which is marked by a front. These two air masses (i.e. air with significantly different temperatures and/or humidities) do not tend to mix easily, because cold air is denser compared to warm air, and the wedge of warm air will be forced to rise over the cold air. When this happens, air pressure falls at the surface and a cyclonic circulation can result.

As the cyclone starts rotating, a cold front and warm front will arise. These mark the boundaries between the two air masses. Areas of cloud, rain or snow form near these weather fronts. The cold front tends to sweep towards the equator and move into the back of the cyclone. The warm front ahead will move more slowly, as it comes up against cold air ahead of the system. Eventually, the cold front overtakes the warm front and all of the warm air is lifted above the surface. When this happens, the temperature contrast is lessened and the cyclone begins to weaken.

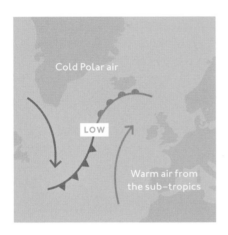
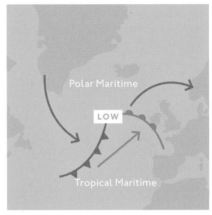
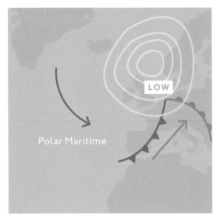
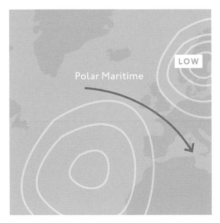

HURRICANES

HURRICANES

TYPHOONS

CYCLONES

EQUATOR

CYCLONES

No Hurricanes Here, Please

The only difference between a cyclone, typhoon or hurricane is where the storm occurs. In the Atlantic Ocean and the north-east Pacific, they are called hurricanes. In the north-west Pacific, they are called typhoons. In the South Pacific and Indian Ocean, they are called cyclones. A hurricane, or tropical cyclone, cannot form near the UK since the sea surface temperatures are much too low. However, although very rare, it is possible for extratropical cyclones to hit the UK with hurricane-force winds. It's also possible for ex-hurricanes to hit the UK, after undergoing a process called extratropical transition.

Normally, hurricanes weaken when encountering the cooler seas of the North Atlantic. But sometimes they can be given a new lease of life if they drift into the midlatitudes and interact with the boundary between cold polar air and warm subtropical air. The very warm air associated with the storm adds to this contrast and a vigorous midlatitude depression can result.

An example of an ex-hurricane influencing weather in the UK was Ophelia, which hit on 16 October 2017, bringing very strong winds to western parts of the UK and Ireland – the top gust speed recorded was 90 miles per hour at Capel Curig and Aberdaron in Wales.

The Storm that Inspired the Shipping Forecast

On 25 October 1859, the steam clipper the *Royal Charter* was sunk during a severe storm in the Irish Sea. The 'Royal Charter Gale', as it became known, took 800 lives and sunk 133 ships, with 90 more vessels badly damaged. Twice as many people were lost at sea around the British Isles during the storm than in the whole of 1858, and its severity inspired Met Office founder Robert FitzRoy to start the world's first national weather-warning service. FitzRoy, founder of the fledgling Meteorological Office, which had been collecting observations from around the British coast since 1854, believed his department could have predicted the storm, and he produced a detailed report with charts to substantiate his claim. Through his analyses of the 'Royal Charter Gale' and other storms, FitzRoy demonstrated the validity of his models and proposed a national storm-warning system.

Many people in the scientific establishment doubted that weather could be predicted in any meaningful way. However, the British government permitted FitzRoy to test his science of weather forecasting and establish his service. In February 1861, the first warning was issued using a combination of cones and drums hoisted on masts to warn vessels both in harbour and along the coast of an approaching gale. On the River Tyne, the warning was disregarded and many lives were lost, but the accuracy of the forecast ensured that further warnings were heeded, and the scheme became widely popular. Supposedly the longest-running national-forecasting service in the world, nowadays it is better known as the iconic Shipping Forecast, provided by the Met Office on behalf of the Maritime and Coastguard Agency.

What Is a Storm Surge?

A storm surge is possibly one of the most dramatic weather events in the UK, resulting from low pressure, high winds and tidal conditions. The primary cause is high winds pushing seawater towards the coast, causing it to pile up there. There is also a smaller contribution from the low pressure at the centre of the storm 'pulling' the water level up, by about one centimetre for every one millibar change in pressure. This is called the inverse barometer effect and is similar to what happens when you drink through a straw. The strong winds in the storm generate large waves on top of the surge that can cause damage to sea defences or spill over the top of them, adding to the flood risk.

The height of a storm surge depends on many factors, including the size and strength of the storm, the direction by which it approaches the coast, and the shape of the coastline and seabed. In areas with large tides, such as the UK, the timing of a storm surge is particularly important, and just a couple of hours earlier or later can prove the difference between an area being flooded or staying safe.

On the night of 31 January 1953, a storm in the North Sea caused a storm surge that occurred at the same time as a high spring tide. Although the storm and surge were forecast in advance, public-warning systems were not very effective at the time and many people were not prepared for the flooding. More than 2,500 people were killed around North Sea coastlines, including 307 in England and 19 in Scotland. As well as the loss of life, the flooding caused a great deal of damage to people's homes and businesses, and ruined large areas of farmland. Following the 1953 storm surge, the UK government invested in improved sea defences, including the Thames Barrier, and much more effective warning systems.

It is also possible for a negative storm surge to occur, when the wind direction instead blows the water away from the coast, causing the sea level there to drop. These are less dangerous than positive storm surges as they do not bring the risk of flooding, but they can damage ships in port and leave them stranded until the water level rises again.

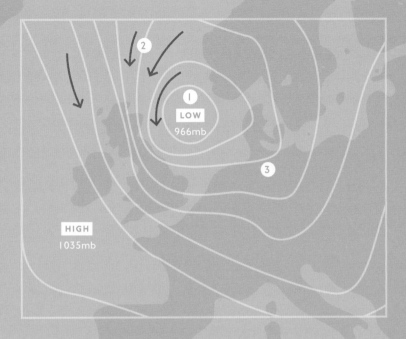

1 Deep low pressure over the North Sea allows sea levels to rise.
2 Strong winds from the north funnel water as it is pushed south.
3 Increased sea levels and strong onshore winds
   causes a storm surge and coastal flooding.

Discovering the Jet Streams

The jet streams were first discovered during the nineteenth century when
meteorologists noticed that high-level cloud (cirrus) was moving very
rapidly across the sky. Then, between 1923 and 1925, Japanese scientist
Wasaburo Oishi recorded unusually high winds by observing balloons
flying quickly away near Mount Fuji. His findings weren't widely known,
however, because he published them in Esperanto. The existence of jet
streams was more widely acknowledged during the Second World War
after US B29 bombers noticed a significant head wind, which caused
them to run low on fuel during a raid above Tokyo. The jet stream is now
a fundamental part of weather forecasting, because of its ability to spawn
areas of low pressure and storms at the surface (see 'Changeable').

The Great Storms of 1703 and 1987

Often cited as the worst on record, the 'Great Storm of 1703' brought widespread disruption to England and Wales and inspired the novelist Daniel Defoe's book *The Storm*, a non-fiction account of the event. Following weeks of wet and windy weather, the extratropical cyclone hit during the middle of the night on 26 November and caused destruction across much of Wales, the Midlands and the south-east, with Bristol and London suffering in particular. It felled 4,000 oak trees in the New Forest alone, destroyed more than 400 windmills, brought down around 2,000 huge chimney stacks in the City of London and even blew the roof off the Palace of Westminster. Such was its impact that Queen Anne was moved to say it was 'a Calamity so Dreadful and Astonishing, that the like hath not been Seen or Felt, in the Memory of any Person Living in this Our Kingdom'.

In more recent times, the 'Great Storm of 1987' brought more extreme weather and destruction to our shores. Winds gusting at up to 100 miles per hour were recorded, and around 15 million trees were blown down. Many fell onto roads and railways, causing major transport delays. Others took down electricity and telephone lines, leaving thousands of homes without power for more than twenty-four hours. Buildings were also damaged by winds or falling trees, and numerous small boats were wrecked or blown away, with one ship at Dover being blown over, and a Channel ferry was blown ashore near Folkestone. While the storm took a human toll, claiming eighteen lives in England, it is thought many more might have been hurt if the storm had hit during the day.

What Is a Weather Bomb?

Weather bomb is another term for explosive cyclogenesis, which is when an area of low pressure deepens so rapidly that its central pressure falls by 24 hPa or more in the space of twenty-four hours. They occur most frequently over the sea in winter, and can produce winds of between 80 and 95 miles per hour.

Unlucky Michael Fish

The 'Great Storm of 1987' has become particularly infamous thanks to TV weather presenter Michael Fish, who told viewers there would be no hurricane on the evening before the storm struck. He was unlucky, however, as he was talking about a different storm system over the western part of the North Atlantic Ocean that day. This storm, he said, would not reach the British Isles – and it didn't. It was a rapidly deepening depression from the Bay of Biscay that struck. This storm wasn't a hurricane, as it did not originate in the tropics and had a different structure, but it was certainly exceptional. In the Beaufort Scale of wind force, which measures how strong the wind is, hurricane force (Force 12) is defined as a wind of sixty-four knots or more, sustained over a period of at least ten minutes. By this definition, Hurricane Force winds occurred locally during the 1987 storm but were not widespread.

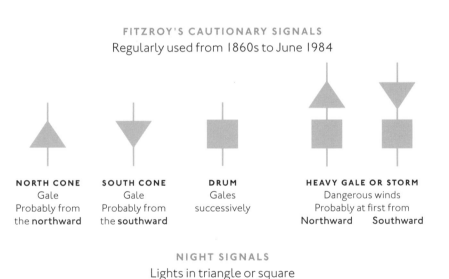

FITZROY'S CAUTIONARY SIGNALS
Regularly used from 1860s to June 1984

NORTH CONE
Gale
Probably from
the **northward**

SOUTH CONE
Gale
Probably from
the **southward**

DRUM
Gales
successively

HEAVY GALE OR STORM
Dangerous winds
Probably at first from
Northward   Southward

NIGHT SIGNALS
Lights in triangle or square

4 lanterns and 2 halyards, each of good rope protected from chafing.
The signals can be made with a lantern of any colour, although white and red
are the most easy to see. The signals should hang at least 3 feet apart.

Weather Warnings Today

The effects of the weather in the UK are extremely varied, both from place to place and from season to season. A 50 miles per hour wind gust, for example, would have a bigger impact in Birmingham in August – when there may be outdoor events taking place, with temporary structures and lots of people in attendance – compared with the same strength of wind in Shetland during January, where it is a relatively frequent occurrence and the populations density is low. That's why Met Office Severe Weather Warnings are based on an impact-likelihood matrix. The higher the impact and the higher the likelihood of that impact, the more serious the warning – in order from yellow to amber to red.

WARNING IMPACT MATRIX

VERY
LIKELY

UNLIKELY

VERY
LOW
IMPACT

HIGH
IMPACT

The Beaufort Scale in Action

This chart shows some of the visual clues you can look for to work out what the Beaufort number might be (*see overleaf*).

BEAUFORT NUMBER	WARNING FLAG	INTENSITY CATEGORY	WIND SPEED	WAVE HEIGHT
0		Calm	< 1 knot < 1mph	0ft
1		Light	1–3 knots 1–3mph	0–1ft
2		Light breeze	4–6 knots 4–7mph	1–2ft
3		Gentle breeze	7–10 knots 8–12mph	2–4ft
4		Moderate breeze	11–16 knots 13–18mph	3.5–6ft
5		Fresh breeze	17–21 knots 19–24mph	6–10ft
6		Strong breeze	22–27 knots 25–31mph	9–13ft
7	▶	High wind, moderate gale, near gale	28–33 knots 32–38mph	13–19ft
8	▶▶	Gale, fresh gale	34–40 knots 39–46mph	18–25ft
9	▶▶	Strong/ severe gale	41–47 knots 47–54mph	23–32ft
10	■	Storm, whole gale	48–55 knots 55–63mph	29–41ft
11	■	Violent storm	56–63 knots 64–72mph	37–52ft
12	■■	Hurricane force	≥ 64 knots ≥ 73mph	≥ 46ft

STORMY

SEA CONDITIONS	LAND CONDITIONS
Sea like a mirror	Smoke rises vertically
Ripples with appearance of scales are formed, no foam crests	Direction shown by smoke drift but not by wind vanes
Small wavelets; crests have a glassy appearance but do not break	Wind felt on face; leaves rustle; wind vane moved by wind
Large wavelets; crests break; foam of glassy appearance	Leaves and small twigs in constant motion; light flags extended
Small waves becoming longer; fairly frequent white horses	Raises dust and loose paper; small branches moved
Moderate waves with pronounced longer form; many white horses are formed	Small trees in leaf begin to sway; crested wavelets form on inland waters
Large waves; white foam crests are more extensive everywhere; probably some spray	Large branches in motion; umbrellas used with difficulty
Sea heaps up and white foam from breaking waves begins to be blown in streaks along the direction of the wind; spindrift begins to be seen	Whole trees in motion; difficulty walking against the wind
Moderately high waves of greater length; edges of crests break into spindrift; foam is blown in well-marked streaks along the direction of the wind	Twigs break off trees; generally impedes progress
High waves; dense streaks of foam along the direction of the wind; sea begins to roll; spray affects visibility	Slight structural damage (chimney pots and roof slates removed)
Very high waves with long overhanging crests; foam blown in dense white streaks along the direction of the wind; on the whole the surface of the sea takes on a white appearance; rolling of the sea becomes heavy; visibility affected	Seldom experienced inland; trees uprooted; considerable structural damage
Exceptionally high waves; small- and medium-sized ships might be for a long time lost to view behind the waves; sea is covered with long white patches of foam; everywhere the edges of the wave crests are blown into foam; visibility affected	Very rarely experienced; accompanied by widespread damage
The air is filled with foam and spray; sea is completely white with driving spray; visibility very seriously affected	Devastation

What's In a Name?

Following the introduction of the 'Name Our Storms' initiative in 2015, storms are now named using a single authoritative system to increase awareness of the potential impacts of forecast severe weather, allowing people to take action and prepare themselves and their loved ones. A storm will be named by the Met Office when it has the potential to cause an amber or red warning. Other weather types will also be considered, specifically rain if its impact could lead to flooding. Storm systems can, therefore, be named on the basis of impacts from the wind but also include the impacts of rain and snow.

The Met Office work with Met Éireann and KNMI (the Irish and Dutch national meteorological services) to provide a consistent message to people living in and travelling across the three countries. Names are suggested by the public, and the most popular are taken through and make up a list for the year to come, which runs from September to August. The naming convention follows the established method used by the National Hurricane Center in the US, which starts at the beginning of the alphabet for the first storm that qualifies in any given year and then makes its way down the list, although it does not include the letters Q, U, X, Y or Z. Names are compiled from the UK, Ireland and Netherlands to give a good representation of each country and its population.

As of 2020, there have been thirty-six named storms, with the stormiest season so far happening in 2015–16. That year, there were eleven named storms, getting as far as Katie on the list. Katie was also one of only four named storms to see gust speeds in excess of 100 miles per hour; the others were Gertrude, Eleanor and Ali.

How to Name a Storm

If you want to get involved, suggest names for future storms by emailing nameourstorms@metoffice.gov.uk and following Met Office's social-media channels for more information.

A note: we tend to steer clear of names that relate to politicians or the Royal family!

STORMS OF 2015–16

Abigail (12–13 Nov 2015)
Barney (17–18 Nov 2015)
Clodagh (29 Nov 2015)
Desmond (4–6 Dec 2015)
Eva (23–24 Dec 2015)
Frank (29–30 Dec 2015)
Gertrude (29 Jan 2016)
Henry (1–2 Feb 2016)
Imogen (7–8 Feb 2016)
Jake (2 Mar 2016)
Katie (28 Mar 2016)

STORMS OF 2016–17

Angus (19–22 Nov 2016)
Barbara (22–27 Dec 2016)
Conor (23–29 Dec 2016)
Doris (21–26 Feb 2017)
Ewan (25 February–3 Mar 2017)

STORMS OF 2017–18

Aileen (12–13 Sept 2017)
Brian (21 Oct 2017)
Caroline (7–10 Dec 2017)
Dylan (30–31 Dec 2017)
Eleanor (2–3 Jan 2018)
Fionn (16 Jan 2018)
Georgina (23–24 Jan 2018)
Hector (13–14 Jun 2018)

STORMS OF 2018–19

Ali (19 Sept 2018)
Bronagh (20–21 Sept 2018)
Callum (12–13 Oct 2018)
Deirdre (15–16 Dec 2018)
Erik (8–9 Feb 2019)
Freya (3–4 Mar 2019)
Gareth (12–13 Mar 2019)
Hannah (26–27 Apr 2019)

STORMS OF 2019–20

Atiyah (4–12 December 2019)
Brendan (11–18 January 2020)
Ciara (7–16 February 2020)
Dennis (11–18 February 2020)

freezing [free-zing]

adjective turning to ice; extremely cold

origins: before 1000; Old English *frēosan*, Middle English *fresen*; cognate with Middle Low German *vrēsen*, Old Norse *frjōsa*, Old High German *friosan* (German *frieren*)

FREEZING

Freezing conditions are when the temperature is at or below 0°C, but more generally 'freezing' is a word we use to describe any cold weather in Britain.

Also Known As...

BALTIC	FROSTY
BITING	ICY
BITTER	NIPPY
BRASS MONKEYS	PARKY
CHILLY	PIERCING
CRISP	WINTRY

As with much of the UK's weather, cold conditions are usually determined by high or low pressure.

What Causes Cold Weather in the UK?

In the winter, under prolonged spells of high pressure known as blocks, when the conditions remain relatively stable, temperatures can drop further and further as cold air is not replaced by milder air, usually from the west. When high pressure becomes established over a large area, it can deflect low-pressure systems, further prolonging the settled but cold conditions.

During early spring and winter, the sun does not have enough energy to warm the ground – because of the limited sunshine hours and the greater distance the sun's energy has to travel through the atmosphere of the Earth – and the stored heat therefore seeps out into the atmosphere, causing the temperature to drop even further.

Strong winds from the north driven by low pressure can bring very cold conditions and often snow to many parts of the UK, particularly the north. This once again comes down to air masses. In the winter, after a low-pressure area has moved through, the winds can pull down a very cold pool of air and drag an Arctic-maritime air mass across the British Isles.

The North Atlantic Oscillation

There are a number of global factors that contribute to the conditions that we can expect to see in the UK over the winter months, including the North Atlantic Oscillation (NAO). At its simplest, this is the year-to-year change in the direction of the winds over the North Atlantic region. In what is known as its positive phase, lower than normal sea-level pressure occurs near Iceland and higher than normal pressure over the Azores. This enhanced contrast in pressure strengthens the westerly winds from the Atlantic, bringing milder maritime air to the UK but also increased rainfall and more windstorms. Conversely, in a negative NAO phase, the pressure is higher than usual over Iceland and lower over the Azores, weakening the pressure gradient and reducing the strength of westerly winds. For the UK, this means that more of our weather will come from the north or east, which in winter results in colder, drier and potentially snowy conditions.

Why Doesn't the Sea Freeze?

The water temperature needs to fall to around -1.8°C for the sea to freeze, but the lowest it usually gets to in the UK is between about 6 and 10°C. In addition, large bodies of water hold their warmth longer than land and the sea is constantly in motion, making it harder to freeze. But although the seas don't freeze very often around the coast of the UK, that is not to say that they can't. During the freezing winter of 1962–63, the coldest on record for more than 200 years, sea water froze in some of England's harbours, and ice patches formed at sea and on beaches.

BRITAIN'S
LONGEST ICICLE

8.2m in 2010
in Grantown,
Inverness-shire

The Tropical Effect

Although it may seem unlikely, tropical rainfall, which is closely linked to underlying ocean temperatures, can also influence whether we experience cold conditions in the UK. Ocean temperatures change over weeks and months, much more slowly than our day-to-day weather. As a result, they can have big swings from one year to the next, due to events such as an El Niño (when sea temperatures in the tropical eastern Pacific rise 0.5°C above the long-term average), while remaining relatively constant within an individual winter

When an El Niño occurs, impacts are felt far and wide across the globe. Drier weather affects Australia, Indonesia and South Africa while wetter conditions are experienced in Chile, Peru and parts of the US. Compared to the rest of the world, impacts are relatively weak in the UK but it is thought that El Niño can result in a cold end to the winter.

Another phenomenon known to affect UK winter weather is the Madden-Julian Oscillation (MJO). This is the name given to a periodic enhancement of tropical rainfall near the equator that circles the world every one to two months. When the MJO leads to a particularly energetic flare-up of thunderstorms, impacts can be felt far and wide. This happened in January 2018 when a huge cluster of thunderstorms disturbed the atmosphere above the tropical west Pacific. The shockwaves from these storms reverberated throughout the global atmosphere, ultimately helping to destabilise the Stratospheric Polar Vortex and lead to the Beast from the East in March (*see overleaf*).

The Gulf Stream and North Atlantic Current

What does Glasgow, with its average January temperature of 4.4°C, have in common with Novosibirsk, the largest city in Siberia, with its average January temperature of -16.5°C?

The answer is they both sit at roughly the same latitude. So why the large discrepancy in average temperatures? Of course, there are many factors that explain the different winter weather conditions in the cities, but two of the most significant are the Gulf Stream and the North Atlantic Current. These are warm-ocean currents that transport heat from the Gulf of Mexico towards north-west Europe. They are part of the much larger Atlantic Meridional Overturning Circulation, which is a major oceanic current that acts as a thermostat for many of the world's climates. Without this thermostat, the UK's climate would be much colder – it's thought by as much as 3 or 4°C.

In the figure, the labels read:

- LABRADOR CURRENT
- NORWEGIAN CURRENT
- NORTH ATLANTIC DRIFT
- GULF STREAM
- NORTH ATLANTIC SUBTROPICAL GYRE
- CANARY CURRENT
- NORTH EQUATORIAL CURRENT

Beast from the East

'Beast from the East' is a pretty catchy phrase used to describe cold and wintry conditions in the UK as a result of easterly winds from the near Continent. When pressure is high over Scandinavia, the UK tends to experience a polar-continental air mass. When this happens in winter, cold air is drawn in from the Eurasian landmass, bringing the cold and wintry conditions that give rise to the name.

The type of weather this cold air mass brings with it depends on the length of time it spends over the sea, or its sea track, during its passage from Europe to the British Isles. The polar-continental air mass is inherently very cold and dry. If it reaches southern Britain after a short sea track over the English Channel, the weather is characterised by clear skies and severe frost. With a longer sea track over the North Sea, the air becomes unstable and moisture is added, giving rise to showers of rain or snow, especially near the east coast of Britain. Polar-continental air only ever reaches Britain between November and April, and the UK's lowest temperatures usually occur in this air mass: lower than -10°C at night, and sometimes remaining below freezing all day.

The most significant 'Beast from the East' in recent years was in 2018 when a large area of high pressure dominating Scandinavia and northern Europe resulted in an easterly airflow across the UK, drawing bitterly cold air from Finland, north-west Russia and the Barents Sea. This situation was associated with a stratospheric warming event over the previous few days (*see overleaf*), blocking the jet stream and milder air associated with Atlantic weather systems. Freezing temperatures combined with a strong east wind, particularly on 28 February and 1 March, resulting in a wind chill at times widely below -10 °C.

This was the most significant spell of snow and low temperatures for the UK overall since December 2010, and it brought widespread travel chaos. Many roads were closed, rail services cancelled, air transport disrupted and cars stranded overnight in both Scotland and England. Thousands of schools across England, Wales and Scotland were closed, and many areas suffered power cuts, while isolated communities and farms across the North Pennines had supplies delivered by helicopter. Beastly!

The Highest Weather Observatory in Britain

Between 1883 and 1904 there was a meteorological observatory on top of Ben Nevis, Britain's highest mountain. The people who operated it lived there all year round, and the average recorded temperature was freezing, although it was often far colder and extremely windy.

BRITAIN'S HIGHEST OBSERVATORY

1,345m in 1883–1904 on Ben Nevis, Scotland

Weather from the Stratosphere

In recent years, some extreme cold and winter snow events have been connected to the surface effects of sudden stratospheric warmings, such as those in 2009–10, 2013 and the 2018 'Beast from the East'. So why, then, is it called stratospheric warming if it leads to cold weather at surface level?

Sudden stratospheric warming refers to a rapid warming of up to about 50°C in just a couple of days, between 10 kilometres and 50 kilometres above the Earth's surface. This is so high up that we don't feel the 'warming' ourselves. However, we start to see knock-on effects on the jet stream, which in turn affects our weather lower down in the troposphere.

Every year in winter, strong winds of up to 155 miles per hour – the strength of the winds in the strongest hurricanes (*see* 'Stormy'), known as Category 5 – circle around the pole high up in the stratosphere. This is called the polar vortex, and it circulates from west to east around cold air high over the Arctic. In some years, polar vortex winds temporarily weaken, or even reverse to flow from east to west. The cold air then descends very rapidly in the polar vortex, and this causes the stratosphere's temperature to rise very quickly. As the cold air from high up in the stratosphere disperses and sinks into the troposphere, it can affect the shape of the jet stream. It is this variation that causes our weather to change.

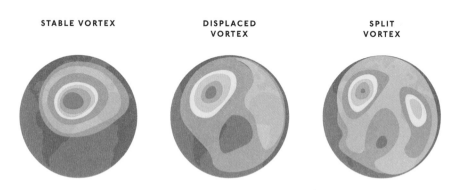

STABLE VORTEX DISPLACED VORTEX SPLIT VORTEX

The Polar Vortex is a circulation of winds high up in the stratosphere (50km above the Earth) that strengthens and weakens.

Sudden stratospheric warming can sometimes cause the jet stream to 'snake' more, which tends to create a large area of persistent high pressure. Typically, this will form over the North Atlantic and Scandinavia. This means that northern Europe, including the UK, is likely to get a long spell of dry, cold weather, whereas southern Europe will tend to be more mild, wet and windy. On the boundary of these areas, cold easterly winds develop, and in some cases the drops in temperature lead to snow, which is what happened in early 2018.

We can reliably predict individual sudden stratospheric warmings about a week in advance using satellites and other observations. This means we have some time to see how they might develop and impact our future weather, particularly as the sudden stratospheric warming usually takes a few weeks to influence conditions at the surface.

Wind Chill

The temperature doesn't always have to reach zero for us to say we're freezing. One of the main factors in how cold it feels to us is wind chill, which could also be described as the 'feels like' temperature – this is different from the actual air temperature shown on a weather forecast.

There is no official definition of wind chill, and the way it is measured varies around the globe. In the UK, a system called the Joint Action Group for Temperature Indices is used. The 'feels like' temperature measures the expected air temperature, relative humidity and the strength of the wind at 1.5 metres (human height), along with a formula that provides an understanding of how heat is lost from a person's bare face at a walking speed of 3 miles per hour during cold and windy days.

The 'feels like' temperature is especially important on windy days, due to the effect of wind on the evaporation speed of moisture from skin: the stronger the wind, the faster the cooling of the skin. On a calm day, our bodies insulate us with a boundary layer, which warms the air closest to the skin. If it is windy, the wind will take the boundary layer away and the skin temperature will drop, making us feel colder.

The Little Ice Age

Following a period of generally warmer weather during the early medieval times, the global weather became colder on average in what is now known as the Little Ice Age. This was not a true ice age, but summers in the northern hemisphere were no longer guaranteed to be warm, and winters could be harsher than we experience today. There are different ways to define this period, although it is generally thought to have lasted from about the fourteenth to the end of the nineteenth centuries.

The Thames froze on several occasions during the Little Ice Age, and the ice was thick enough for Frost Fairs to be held on the river. The first fair was in 1608 and the last in 1814, and they consisted of food and drinks stalls, and a wide variety of side shows and entertainments. Horses and carriages were known to travel on the ice, and in one particularly daring stunt, during the final fair, an elephant was even led across the river.

The river was wider and slower at this time, which allowed it to freeze over when temperatures dropped below 0°C. The old London Bridge, before it was removed in 1835 (its replacement having been opened in 1831), also trapped debris, which slowed the water flow even further and allowed it to ice over. The river froze 24 times between 1400 and 1835.

Years the River Thames Froze since 1400

	FIFTEENTH CENTURY	SIXTEENTH CENTURY
		1506
		1514
		1537
	1408	1565
	1435	1595

Frost Hollows

Frost hollows are another example of a British microclimate, this time ones that can lead to freezing conditions in isolated locations. In the countryside, overnight temperatures – especially during clear and calm nights in the winter – can be significantly lower than in cities. This is especially the case in so-called frost hollows. A frost hollow is a localised dip or depression in the surface of the Earth in which dense cold air sinks and accumulates on a calm night. This cold air drainage results in lower overnight temperatures within the frost hollow compared to its surroundings. One example of a frost hollow is RAF Benson in Oxfordshire.

Frost hollows can also be enhanced by their soil type. Sandy soils contain more pockets of air within them and less moisture compared to damper, heavier clay soils. This helps them cool more quickly at night and warm up faster during the day. Santon Downham in Norfolk, for example, has sandy soils and is often three degrees colder compared to its surroundings on a calm, clear winter's night.

SEVENTEENTH CENTURY

1608
1621
1635
1649
1655
1663
1666
1677
1684
1695

EIGHTEENTH CENTURY

1709
1716
1740
1776
1788
1795

NINETEENTH CENTURY

1814

Ice, Glorious Ice

Ice is not just something you drop in your drink – it comes in many shapes and sizes, even in the more temperate climate of the UK.

1 **BLACK ICE** Glaze ice (or clear ice) that forms on roads and pathways is often called 'black ice' due to its transparent nature, allowing the road surface below to be seen through it. Black ice is particularly dangerous as it can appear almost invisible to drivers, and to unsuspecting pedestrians.

2 **GROUND FROST** A ground frost refers to the formation of ice on the ground or other objects such as trees that have surface temperatures below freezing point. When the ground cools quicker than the air, a ground frost can occur without an air frost (air temperature below freezing at a height of at least one metre above the ground).

3 **HAIR ICE** Hair ice is a rare type of ice formation that occurs when the presence of a fungus called *Exidiopsis effuse* in rotting wood produces thin strands of ice that resemble hair or candy floss.

4 **HOAR FROST** Hoar frost occurs in the same way as dew but below 0°C. A 'feathery' variety forms when the surface temperature reaches freezing point before dew begins to form on it. A 'white' frost, composed of more globular ice, occurs when the dew forms first then freezes.

5 **ICE PANCAKES** Ice pancakes are 20- to 200-centimetre discs of ice that form on bodies of water. Although ice pancakes look solid, they are often quite slushy and easily break apart when lifted up.

6 **ICE PELLETS** Ice pellets form when snowflakes start to melt as they drop from a cloud, then fall through sub-freezing air, where they refreeze into grain-like particles. Ice pellets are generally smaller than hailstones and bounce when they hit the ground.

7 ICE SPIKES Ice spikes form as a result of water freezing from the outside in. If a weakness or hole appears in the 'skin' of the ice, the liquid water inside is squeezed up through it, freezing as it goes.

8 ICICLES Icicles form when ice or snow melts and refreezes in long spikes towards the ground. Their ends can often be sharp and therefore dangerous if they snap and fall.

9 RIME ICE Rime is a rough white ice deposit that forms on vertical surfaces exposed to the wind. It is formed by supercooled water droplets of fog freezing on contact with a surface as it drifts past.

The Rare and Dangerous Phenomenon of Freezing Rain

Freezing rain is a rare type of liquid precipitation that strikes a cold surface and freezes almost instantly. When precipitation that falls from a cloud as snow descends through warmer air before reaching the ground, it melts and turns into rain droplets. On rare occasions, if it then falls through cold air again just before hitting the earth, the droplets can become 'supercooled'. This means that they are still falling in liquid form, even though their temperature has fallen below 0°C. When this supercooled droplet hits the ground (which is below 0°C too), it spreads out a little on landing, and then instantly freezes, encasing the surface in a layer of clear ice.

Freezing rain can produce striking effects. It is also extremely dangerous, making all exposed surfaces very slippery. And on some occasions, freezing rain can turn to ice on contact with trees and power cables, and if heavy enough, it has the potential to bring them down.

Thankfully, we don't see this phenomenon very often in the UK, but it does occur from time to time. On the evening of 23 January 1996, an area of low pressure moved slowly northwards across the south of England, leading to outbreaks of rain. In some areas, the rain turned to snow, but in parts of Wales, south-west England and the Midlands the snow turned to freezing rain, with outbreaks through the night and intermittently through 24 January.

The most significant impact of this freezing-rain event was a spate of road accidents, and in the Birmingham area a 50 per cent increase in hospital admissions was reported. As well as car accidents, people slipped on icy pavements. So, make sure you keep your wits about you when this supercooled phenomenon occurs.

BRITAINS LOWEST TEMPERATURE

–27.2°C on 10 January 1982; 11 February 1895 at Braemar; and 30 December 1995 at Altnaharra.

Britain's Longest Icicle

Although the lengths of icicles are not officially recorded, one discovered on a bridge over the River Spey in Grantown, Inverness-shire, must have a good claim to being Britain's longest ever. It was thought that water from a leaking pipe in the bridge was the source of the mammoth icicles, the longest of which grew to approximately 8.2 metres.

What Is Absolute Zero?

The most commonly used measurement of temperature is the Celsius scale, which was designed to reference the freezing point (0°C) and the boiling point (100°C) of water. There is no upper limit, but the lowest temperature possible is -273.15°C: absolute zero. At this temperature, all the particles in a substance stop moving. They have no energy left to lose, so the substance cannot get any colder.

Many scientists use the Kelvin scale instead, which is also more useful in space, where there is no water. Kelvin begins at absolute zero, so there are no minus numbers, which makes calculations simpler. An increase of one degree Celsius is the same as one kelvin.

The Albedo Effect

Albedo is a measure of how reflective a surface is, specifically the proportion of the incoming solar radiation that is reflected by the surface of the Earth back into the atmosphere. The term is derived from the Latin *albus*, meaning 'white', and it is determined either by a value between 0 and 1 or as a percentage. The more reflective a surface is, the higher the albedo value. (Next time you look in a mirror, feel free to remark upon its great albedo credentials.) Very white surfaces, such as fresh snow, particularly at the poles, reflect a very high fraction of incoming radiation back to space. Darker surfaces such as water, forests or asphalt have a much lower albedo and more of the sun's energy is therefore absorbed.

Albedo is a very important factor in both weather and climate. As different surface types absorb different amounts of energy, heating to differing extents, temperature gradients are created in the atmosphere above these surfaces and these gradients drive our weather systems. Changes in the albedo of the Earth's surface, caused by changes in land surface type, also affect its 'energy budget'. If less energy is reflected, there will be a warming effect. This can lead to feedbacks, as if there is extensive snow melt, more energy will be absorbed, leading to more surface heating and more snow melt. On a large scale, this has major implications for our climate. This is why melting polar ice is of such concern to climate scientists.

snowy [snoh-ee]

adjective abounding in or covered with snow;
characterised by snow, as the weather

origins: before 1000; Middle English *snawy*, Old English *snāwig*

SNOWY

Although not one of the UK's most frequent types of weather, snow is certainly one of its most striking, and potentially one of its most disruptive. It can also be one of the most fun – after all, who doesn't like to build a snowman?

What Is Snow?

Snow is a type of precipitation that occurs when temperatures are low and moisture in the atmosphere freezes to form tiny ice crystals. If enough crystals stick together, they form snowflakes that are sufficiently heavy to overcome any updrafts in the cloud and fall to the ground. Snowflakes that descend through moist air that is slightly warmer than 0°C melt around the edges and stick together to produce bigger flakes, whereas snowflakes that fall through cold, dry air produce powdery snow that does not stick together. This 'dry' snow is ideal for snow sports but is more likely to drift in windy weather. 'Wet' snow, on the other hand, is good for making snowmen and snowballs.

How to Build the Best Snowman

So, wet snow is best for building a snowman, but surely all snow is made of water, so how can it be wetter or drier? One way to classify how wet snow is is not by how much water would be produced if it melted but instead by how much free water it contains relative to the number of ice crystals – the less free water, the drier the snow:

WATER CONTENT IN SNOW

◌ **DRY:** 0 per cent water

💧 **MOIST:** less than 3 per cent water

💧💧💧 **WET:** 3 to 8 per cent water

💧💧💧💧 **VERY WET:** 8 to 15 per cent water

💧💧💧💧💧 **SLUSH:** more than 15 per cent water

Dry snow doesn't stick together as well, and slush doesn't hold its shape, so moist to wet snow is best for snowman building. When it is colder, more water vapour freezes into ice crystals and produces drier snow, meaning that temperatures around or just above 0°C at ground level are ideal. When actually constructing your snowman, it is also helpful to pack the snow as tightly as possible, as this causes it to melt slightly and then refreeze into ice crystals, binding it together. The bigger your snowman, the more difficult it is to pack tightly and the less stable the structure will be. It is quite difficult to make a giant snowman without it collapsing. Spheres are the easiest shape to compact, hence the classic shape of balls of snow decreasing in size, with the largest at the bottom and the smallest at the top. After that, you can let your imagination run wild … although no snowman is complete without a carrot for a nose, of course.

How Cold Does It Need to Be to Snow?

Precipitation can fall as snow when the air temperature is below 2°C, and it is therefore a myth that it needs to be below zero to snow. In fact, in the UK, the heaviest snowfalls tend to occur when the air temperature is between 0°C and 2°C. The air temperature has to be below freezing in the clouds for ice crystals and snow to form in the first place, but snow will often fall through air closer to the surface that is above freezing. If the temperature is warmer than 2°C, the snowflakes will usually melt and fall as sleet rather than snow, and if it's warmer still, it will be rain.

This melting isn't always instant, though. If the temperature at the surface is 1°C or 2°C, but it's below freezing at a height of around 200 metres above ground, then the snow often doesn't melt. This is especially the case when the snow is heavy. Snow falling through the atmosphere also helps to cool the surrounding air by evaporative cooling. It's similar to the effect we all feel when we get out of the shower and the water droplets on our skin evaporate, making us feel colder. Snow falling through the air evaporates slightly, turning back to water vapour. This evaporation requires latent heat energy, which it extracts from the surrounding air, helping to lower the temperature. The heavier the precipitation, the more the evaporation, and the more the air cools down.

Falling rain can also have the same effect on the air temperature. If it starts to rain at around 4°C, as it often can in the UK, especially during the winter, and becomes heavier, the temperature will gradually fall and the snow will penetrate to a lower and lower level until it reaches the ground. It's one of the reasons why snow is so difficult to predict in this country. It can depend not just on initial temperature but also on the height of the freezing layer in the atmosphere, and the intensity and duration of the precipitation. A passing rain shower in one location can become sleet and then snow in another.

Can It Be Too Cold to Snow?

You sometimes hear people say that it is too cold to snow, but can that actually ever be the case? Strictly speaking, the answer is no, but it can be too dry to snow, and the colder air is, the drier it tends to be because cold air holds less water vapour than warmer air. At -15°C, air's capacity to hold moisture is only 25 per cent that of air at freezing point. But even the coldest air observed near the Earth's surface can hold some moisture. So, in theory it can't be too cold to snow, but the colder the air is, the less likely it is that ice crystals will precipitate out and form into flakes to fall to Earth as snow.

This is why major snowstorms at below approximately -10°C are rare. We seldom see temperatures that low in the UK, meaning that it's never too cold to snow in the temperature range we typically experience. Having said that, we do tend to get our lowest temperatures in the UK when the air is dry and the skies are clear overnight. Therefore, when we experience our lowest temperatures, it probably won't be snowing at the same time.

For snow to fall, you need a moisture source – normally from the sea or ocean – and that source of weather will inherently be less cold. Our heaviest snowfalls in the UK therefore occur when freezing air is in place and then warmer air, containing weather fronts and rainclouds, tries to move in from a much milder direction. This warmer air – from the south-west, say – can contain a lot of moisture. As it clashes with the colder, drier air already resident across the UK, you'll find heavy snow at the boundary between these contrasting weather systems.

What Is Sleet?

In the UK, sleet is partially melted snow, or a mixture of rain and snow. This is not to be confused with the American definition, which describes sleet as refrozen melted snowflakes or frozen raindrops. In the UK, we call these ice pellets.

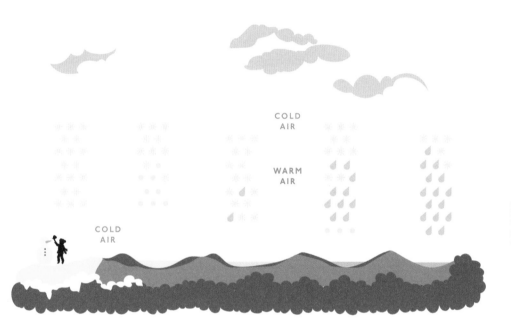

COLD
AIR

WARM
AIR

COLD
AIR

SNOW	**ICE PELLETS**	**SLEET**	**FREEZING RAIN**	**RAIN**
The snow does not melt and falls to the ground as snowflakes	Snow partially melts as it falls briefly in the warm air. It then refreezes into ice pellets as it falls through freezing air closer to the surface	Snow partially melts as it falls briefly in the warm air. It falls as a mix of snowflakes and raindrops to the ground	Snow melts in the warm air, then as the droplets reach the cold air close to the ground they become 'supercooled'. On hitting the ground, the droplet turns to ice.	Snow melts in the warm air and falls to the ground as rain.

What Makes a Snowdrift?

A snowdrift is created when snow is piled up as the wind blows across it. Dry, powdery snow is the easiest type to move around, because it doesn't stick together like wet snow, and the wind speed must be at least 10 miles per hour to get it on the move. The stronger the wind, the more it can shift – it is possible to end up with really deep areas of snow piled against roadsides, houses or any object that's facing the wind.

Loose snow that is lifted into the air by the wind at a height of 2 metres or more is known as 'blowing snow', whereas snow that is lifted into the air by the wind at a height of less than 2 metres is called 'drifting snow'.

The Deepest Snow Ever Recorded in the UK

During the severe winter of 1946–47, a series of cold spells brought large drifts of snow across the UK, causing transport problems and fuel shortages. The deepest snow in an inhabited area of the UK was recorded at this time, with a level-snow depth of 1.65 metres near Ruthin in North Wales in March 1947.

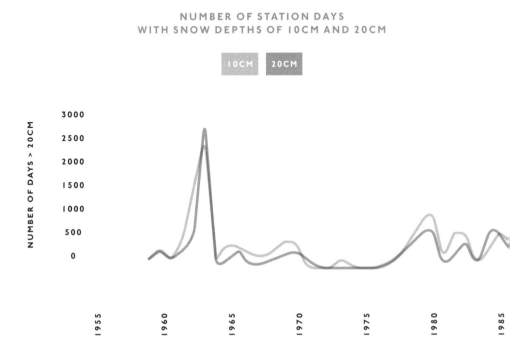

NUMBER OF STATION DAYS
WITH SNOW DEPTHS OF 10CM AND 20CM

'The Wrong Type of Snow'

Railways in the UK seem to fall victim to the vagaries of the very British weather more often than almost anything else. These challenges were perhaps typified by the heavy snowfall in February 1991, which caused major disruption to rail services for more than a week. In this case, the snow had been forecast in advance, and British Rail claimed to be ready for it. However, the snow that fell was drier than expected, making it too deep to be cleared by snowploughs – much less efficient snowblowers were needed instead. As well as blocking the tracks, the soft, powdery snow played havoc with the trains' electrical systems, from motors to sliding doors.

Memories of this event would have probably faded with time if it were not for a now infamous interview with Terry Worrall, British Rail's Director of Operations, on BBC Radio 4's *Today* programme. Worrall explained that 'we are having particular problems with the type of snow, which is rare in the UK'. Interviewer James Naughtie seized on this, saying, 'Oh, I see, it was the wrong type of snow', to which Worall replied, 'No, it was a different type of snow.' The phrase stuck, and it is now often rolled out when there are problems on the railways, as well as being associated with a general sense of disappointment in the apparent inability of British infrastructure to cope with extreme weather events.

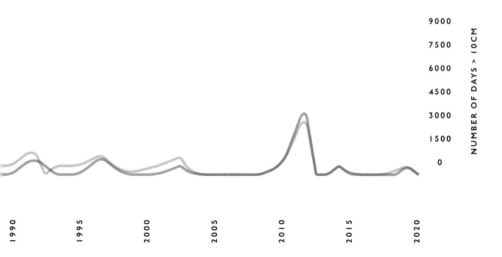

The Heavyweight Snow Championships

JANUARY TO MARCH 1947: there was continuous snow cover from 22 January to 17 March, and snowdrifts 7 metres deep were recorded in the Scottish Highlands.

DECEMBER 1962 TO FEBRUARY 1963: the UK's coldest winter on record, snowdrifts 6 metres deep were recorded in south-west England and Wales, blocking roads and railways, and 23 centimetres of level snow was measured in Manchester city centre.

NOVEMBER 1978 TO MARCH 1979: known as the 'Winter of Discontent', 1.8 to 2.1-metre snowdrifts were recorded on the east coast of England and 4.5-metre snowdrifts in north-east England during March.

APRIL 1981: a very late but intense snow event that saw 23 to almost 27 centimetres of level snow in the Peak District and the accumulation of 6-metre snowdrifts because of strong winds. The subsequent rapid thaw led to major flooding.

DECEMBER 1981 TO JANUARY 1982: the 'Big Snow of 1982' saw 1.8-metre snowdrifts recorded in north-east England.

FEBRUARY 1991: the 'Wrong Type of Snow' (*see previous page*), with 20 centimetres of level snow in London, and villages in Exmoor cut off by 1.8 to 2.1-metre drifts. It also caused chaos on the railways.

DECEMBER 2009 TO FEBRUARY 2010: widespread snow cover across the UK, with 61 centimetres of snow recorded in Aviemore and a 4.5-metre snowdrift recorded in Allenheads in the Pennines.

DECEMBER 2010: the coldest December in more than 100 years, with 76 centimetres of snow recorded in the Peak District.

MARCH 2013: an unusually cold and snowy March, with the UK farming industry badly hit. Forty centimetre-deep snow was reported in Northern Ireland with some areas affected by severe drifts in excess of 2 metres.

MARCH 2018: the 'Beast from the East', with 49 centimetres recorded at St Athan and Drumalbin, and snowdrifts of 7.6 metres in Cumbria.

What Colour Is Snow?

Nil points if you think it's white. Or perhaps half a point.

While snowflakes appear white as they fall through the sky or as they accumulate on the ground as snowfall, they are actually totally clear. The ice crystals, though, are not transparent like a sheet of glass; rather they are translucent, meaning light only passes through indirectly. The many sides of the ice crystals cause a diffuse reflection of the whole light spectrum, which results in snowflakes appearing to be white in colour.

What Are Snow Grains?

Snow grains are the frozen equivalent of drizzle. They generally fall from low-layer clouds such as stratus when the temperature is around or below freezing. They appear as tiny grains of ice and are brittle to the touch.

How Fast Is Snow?

Most snow falls at a speed of between 1 to 4 miles per hour, depending on the individual snowflake's mass and surface area, as well as the environmental conditions it encounters on its descent. Snowflakes that collect supercooled water as they fall can descend at up to 8.7 miles per hour, but snowflakes, as most people recognise them, will tend to float down at around 1.5 miles per hour, taking about an hour on average to reach the ground.

Scotland Wins Snowiest Place in the UK

Snow or sleet falls on 38.1 days per year on average here. The weather station that records the most snowfall in the UK is the Cairngorm Chairlift, with snow falling on average on 76.2 days throughout the year between 1981 and 2010. Perhaps not surprisingly, Cornwall is the place in the UK least likely to get snow, with an average of only 7.4 days of snow or sleet falling a year.

Much of this snowfall does not settle, or lie, on the ground. On average across the UK, there are only 15.6 days a year when snow is on the ground, compared to 26.2 days in Scotland. Most of the snow on the ground can be found in mountainous areas.

DAYS OF SNOWFALL

DAYS* (AVG.)	ELEVATION (M ABOVE SEA)	LOCATION
76.2	663	Cairngorm Chairlift (Cairngorms National Park)
64.7	24	Baltasound (Shetland Islands)
62.8	57	Fair Isle (Shetland Islands)
59.0	28	Loch of Hundland (Orkney Islands)
52.7	253	Copley (County Durham)
51.6	393	Leadhills (South Lanarkshire)
50.2	513	Widdybank Fell (North Pennines)
49.8	236	Eskdalemuir (Dumfries and Galloway)
49.2	103	Kinbrace (Sutherland)
48.5	244	Knockan (Sutherland)

*Annual average days of snowfall between 1981 and 2010

Dreaming of a White Christmas

The concept of a white Christmas is often attributed to Charles Dickens, who grew up during the very cold and snowy winters of the early 1800s. When he wrote *The Pickwick Papers*, which was serialised in 19 editions, he included reference to snowfalls and snow in the scenes set around Christmas. As it happened, these parts of the story were published just days after a very heavy snowfall during Christmas 1836, and the coincidence of fact and fiction aligning created an association between Christmas and snow that has lasted from that point onwards.

Although many of us still love the romanticised notion of a white Christmas today, snow is actually much more likely to fall in January and February than in December. In the UK, snow or sleet falls on an average of 3.9 days in December, compared to 5.3 days in January and 5.6 days in February. It even snows more in March: on average 4.2 days per year. There has been a widespread covering of snow on Christmas Day – judged to be more than 40 per cent of weather stations reporting snow that day – only four times in the last half-century. However, don't lose hope – snow has fallen at least somewhere in the UK on Christmas Day 38 times since 1962.

SNOW ON GROUND AS PERCENTAGE OF ALL REPORTING STATIONS

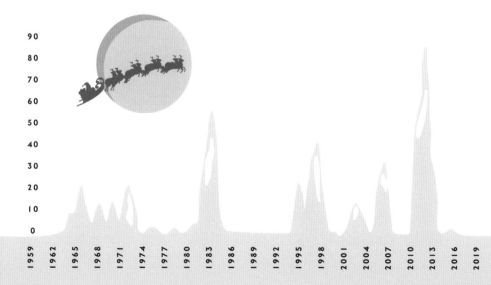

Is Every Snowflake Unique?

It is a commonly held belief that all snowflakes are unique, but as is often the case in nature, the reality is not quite so straightforward. At a molecular level, all snowflakes are by definition distinct, because when you compare the exact type and number of water molecules within even the very smallest ice crystals, it is extremely unlikely that they will ever match. However, at a visual level, some snowflakes do in fact look alike. There are 35 different forms that the ice crystals can take (*see opposite*), depending on the environment in which they were formed. But most snowflakes that descend to Earth do not remain in these simple configurations for long. As the crystals fall through the sky and join together to form larger snowflakes, the combinations are so varied that it is unlikely that a human observer would ever be able to find two exactly the same.

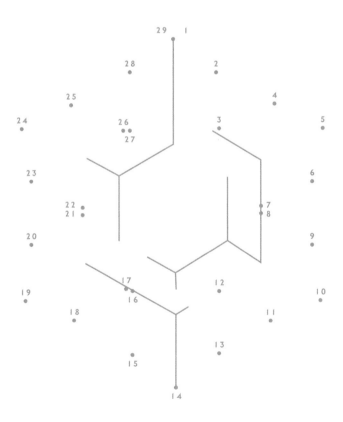

Snow in June

Although the Great British summer can be somewhat unpredictable, and we are not guaranteed sunny weather, it's probably fair to say that most of us don't expect it to snow. But that is what happened on 2 June 1975, when widespread snow was reported in East Anglia, the Midlands and even as far south as London. It was the first time snow had been reported so far south in summer since 1888. Several county cricket matches were abandoned – most notably Essex v Kent at Colchester and Derbyshire v Lancashire at Buxton. With temperatures increasing as a result of climate change, it seems unlikely that we will see a widespread snowfall in the summer months again, but you never know … stranger things have happened when it comes to the weather.

What Is Thundersnow?

Thundersnow is simply the occurrence of thunder and lightning while it's also snowing. It's relatively rare, since thunderstorms form when there is a rapid decrease in temperature with height, and this is more likely to happen when there is warm air close to the surface. However, this temperature contrast can also occur when very cold air arrives high in the atmosphere.

When thundersnow occurs at night, the lightning appears brighter – this is because the light reflects off the snowflakes. The snow contained within the thunderstorm also acts to dampen the sound of the thunder. While the thunder from a typical thunderstorm might be heard many kilometres away, the thunder during a thundersnow event will only be heard if you are within 3 to 5 kilometres of the lightning.

Avalanches

The British weather is often at its most extreme in the mountains, and this is also true when it comes to snow. Avalanches are amongst the most dangerous hazards we are likely to encounter in the UK, especially in the Highlands of Scotland. But what is an avalanche and why do they occur?

In its simplest terms, an avalanche is a moving mass of snow, ice and rock down the side of a mountain. Some are small slides of shapeless, powdery snow that are of limited risk to life and property, but on occasion massive slabs of snow can break free and reach speeds of up to 100 miles per hour, causing destruction in their wake. Most avalanches occur on slopes of 35 to 50 degrees, and an incline of at least 30 degrees is needed for one to occur in the first place, but once they get going and gather momentum they can also travel across flatter ground for short distances.

Avalanches happen because the properties of snow are very sensitive to temperature and humidity. Dry snow is light and accumulations contain a lot of air. A slight rise in temperature and humidity will make the snow heavier. The dry snow is weak, and if a layer of heavier snow builds on top of it, the dry layer can give way allowing a large slab to break free. There are other natural causes of avalanches – earthquakes, for example – but 90 per cent are triggered by humans. Just walking or riding over an area with a heavier layer of snow overlying a weaker one can cause an avalanche to occur. But you don't have to worry about shouting – it is a myth that loud noises can trigger an avalanche.

What Are Blizzards?

The word 'blizzard' originated in North America, the term originally referring to a blast of gunfire. Today, the word has a less violent meaning, although a meteorological blizzard can be just as dangerous. At the Met Office, we now define a blizzard as a moderate or heavy falling of snow, either continuous or in the form of frequent showers, with wind speeds of 30 miles per hour or more and a reasonably extensive snow cover, reducing visibility to 200 metres or less.

Blizzards can occur either when snow is falling in windy conditions or when it is lifted from the ground by strong winds, known as a ground blizzard, or a combination of both. A white out is an extreme form of blizzard in which downdrafts and heavy snowfall combine to create a situation in which it is impossible to tell the ground from the sky or to discern the horizon.

PICTURED: The Big Freeze of 2009-10 that covered the UK in snow.

The UK's Deadliest Avalanche

As snow most often falls in Scotland, it stands to reason that the majority of British avalanches fall in the Highlands, and the Scottish Avalanche Information Service provides an invaluable resource for people spending time in the mountains during the winter. However, the avalanche that claimed the most lives in the UK did not happen in the north of Scotland, as you might expect, but in the south of England.

Lewes in Sussex is only a few kilometres from the south coast, situated on the River Ouse and surrounded by the hills of the South Downs. It was hit by a snowstorm during the particularly bad winter of 1836–37, in which freezing temperatures and heavy snowfalls were common. A blizzard developed on Christmas Eve 1836, leading to large snowdrifts, including a mass of overhanging snow at least 6 metres deep on Cliffe Hill, overlooking the town. A row of workers' cottages were located at the foot of the hill, directly underneath the precariously positioned snow slab.

Despite warnings to vacate their homes, the residents, who were being housed by the local parish and were therefore of limited means, refused to leave, with tragic consequences. The massive snow slab broke free on the morning of 27 December, demolishing the row of houses. Seven people were rescued but eight others lost their lives. Today a pub called the Snowdrop Inn occupies this site.

BRITAIN'S DEADLIEST AVALANCHE

8 deaths, 7 rescued, in December 1836 near Lewes, Sussex

The Inuit Have Nothing on the Scots

Studies have suggested that the Inuit do indeed have many more words for snow than we do in the English language. The dialect spoken in Nunavik, Canada, for example, has at least 53 separate words, including *pukak* (a crystal-like snow that looks like salt), *matsaaruti* (wet snow to ice a sleigh's runners) and *qanik* (falling snow).

However, a project run by the University of Glasgow to create an online Scots thesaurus found that there are more than 400 words or phrases relating to snow in the Scots language, ranging from the straightforward *snaw*, which does not need any further explanation, to the less obvious *scowder* (a thin layer of snow) and *kaavie* (a heavy snowfall or blizzard). And don't forget to put your *scoggers* on if you are venturing out in the snow – a footless legwarmer worn over your *breeks* (trousers).

DAYS OF SLEET/SNOW FALLING ANNUAL AVERAGE 1981–2010

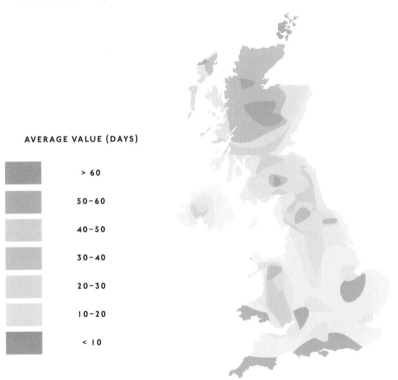

SNOWY

AVERAGE VALUE (DAYS)

> 60

50–60

40–50

30–40

20–30

10–20

< 10

freaky [free-kee]

adjective weird, strange, out of the ordinary

origin: first recorded in 1815–25; origin uncertain, perhaps akin to Old English *frīcian* 'to dance'

FREAKY

Although we experience a wide range of weather conditions in the UK, most do not take us by surprise when they occur, even the less frequent events, such as heavy and widespread snow. On occasion, however, the weather can throw something at us that we've never seen before. From the rare to the downright bizarre, the very British weather therefore still has the power to shock and amaze us from time to time.

What Are Volcanic Winters?

Major volcanic eruptions have a significant effect on the climate. Ash in the upper atmosphere can travel around the globe, blocking the full effect of the sun and significantly reducing temperatures. In extreme cases, this type of event is sometimes referred to as a 'Volcanic Winter'.

One notable example is the eruption of the Icelandic volcano Laki, which began in June 1783 and lasted until February 1784, causing a long-term effect on the climate of the northern hemisphere. Temperatures dropped for two to three years, and ash reached the jet stream and travelled as far as Syria. More locally, gases transported into the atmosphere by the reaction led to falls of acid rain in the UK, with damage to vegetation and crops recorded.

The 2010 Icelandic Eruption

In April 2010, the Icelandic volcano Eyjafjallajökull erupted, bringing with it major disruption to air travel throughout northern and western Europe, with all flights grounded across the UK. Initially lasting for six days, there was also further disruption at times in May of that year, and up to 10 million travellers were affected by the eruption.

A number of factors came together at the same time, resulting in the widespread disruption:

° The volcano erupted directly under the jet stream and was powerful enough to emit ash into it.

° At the time, the jet stream was flowing north-west to south-east from Iceland to north-west Europe.

° The jet stream was in a relatively stable set-up, remaining in the same position for around a week.

° The eruption took place under 200 metres of glacial ice, and the resulting meltwater flowed back into the erupting volcano.

° The meltwater had two knock-on effects: the rapidly vaporising water significantly increased the eruption's explosive power; and the erupting lava cooled very fast, which created a cloud of highly abrasive, glass-rich ash, which was very dangerous for aircraft engines.

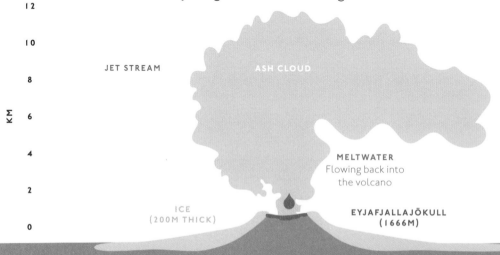

KM

12
10
8
6
4
2
0

JET STREAM   ASH CLOUD

MELTWATER
Flowing back into
the volcano

ICE
(200M THICK)

EYJAFJALLAJÖKULL
(1666M)

0   10   20   30   40

WEST   KM   EAST

Volcanic Sunsets

Volcanic dust in the atmosphere can also create exceptionally brilliant sunsets. Amongst the most famous of these were the sunsets following the eruption of Mount Krakatoa in 1883. Paintings published by the Royal Society show the optical effects created by the 'afterglow' of the eruption in Chelsea, in London, and some people have argued that the sky in Edvard Munch's famous painting *The Scream* is a depiction of what a vivid sunset would have looked like in Norway. Others have more recently argued that Munch was influenced by the appearance of nacreous clouds, which are rare stratospheric clouds known to cause dramatic sunsets.

The Met Office Volcano Watchers

The London Volcanic Ash Advisory Centre (VAAC), hosted and run by the Met Office, is part of the International Civil Aviation Organisation, and one of nine designated centres around the world responsible for issuing advisories for volcanic eruptions. Specialist forecasters use a combination of volcano data, observations from satellites, aircraft and on-the-ground instruments, weather-forecast models, and dispersion models to advise when volcanic eruptions in Iceland are capable of causing disruption to air travel in Europe.

LONDON VOLCANIC ASH ADVISORY CENTRE (VAAC)

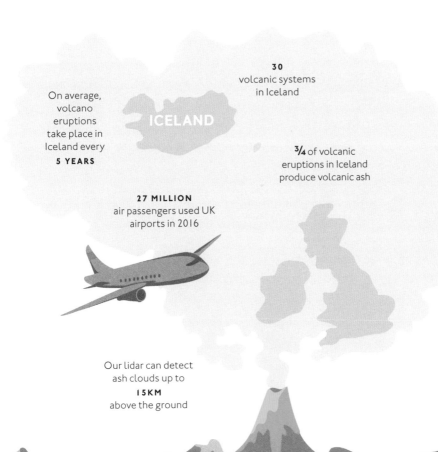

30
volcanic systems
in Iceland

On average,
volcano
eruptions
take place in
Iceland every
5 YEARS

ICELAND

3/4 of volcanic
eruptions in Iceland
produce volcanic ash

27 MILLION
air passengers used UK
airports in 2016

Our lidar can detect
ash clouds up to
15KM
above the ground

The Year Without a Summer

The 'Year Without a Summer' in 1816 followed the eruption of Mount Tambora in Indonesia, in 1815, now considered to be the worst volcanic eruption in human history. Tambora sent 140 billion tonnes of ash into the upper reaches of the atmosphere, with strong stratospheric winds spreading the huge ash cloud across both hemispheres, blocking sunlight in many parts of the world. Terrible weather and lower temperatures led to widespread crop failures and caused food shortages across Europe.

Even before the eruption of Mount Tambora, the period at the end of the eighteenth and beginning of the nineteenth centuries was characterised by unusually cold weather, snowy winters and poor harvests. The 1810s were particularly severe due to a combination of low solar activity as well as increased volcanic activity around the world. But following the Tambora eruption in particular, the UK experienced a very cold spring, with snow falling on 12 May, followed by a wet, cold and occasionally stormy summer.

The British writer Mary Shelley spent the summer on the shore of Lake Geneva (along with Lord Byron, her husband Percy Bysshe Shelley and a couple of others). She wrote:

> ◆It proved a wet, ungenial summer and incessant
> rain often confined us for days to the house ◆.

They amused themselves by setting a challenge to each write a ghost story. Mary's contribution was *Frankenstein*.

What Are Funnel Clouds, Waterspouts and Dust Devils?

Funnel clouds are simply tornadoes that haven't extended fully to the ground. When a funnel cloud hits the ground, it becomes a tornado. However, an observer may not be able to tell the difference if buildings or scenery are in the way of the lower part of the funnel cloud or tornado.

There are two types of waterspouts, spinning columns of air and water mist over water. Tornadic waterspout extend down from a severe thunderstorm either over water or moving from land to sea. They can be dangerous and often accompanied by lightning, hail and strong winds. Fair-weather waterspouts, on the other hand, usually form upwards from the surface of the sea and extend to the base of a developing cumulus cloud – disappearing shortly after reaching the cloud's base. They usually occur when winds are light, so they don't move very far but can still be dangerous.

Tornadoes are a type of whirlwind – the most violent and devastating type – but there are other whirlwinds that form on fine days with light winds. These are wind devils or, more commonly, dust devils, hay devils, fire devils and snow devils, depending on where they occur and the matter enclosed within them. Forming when a rapidly rising column of air encounters a disturbance and begins to rotate, they don't cause much damage or last long.

How Common Are Tornadoes in the UK?

Although most of us have probably never encountered one directly, Britain has more reported tornadoes relative to its land area than any other country except the Netherlands – with an average of 33 reported each year.

According to TORRO, the Tornado and Storm Research Organisation, most reports are from the Midlands, central southern and south-east England, and East Anglia – they are normally associated with large, energetic thunderstorms, more common to these areas. Tornadoes occur occasionally in the south-west, north-west and north-east of England and in Wales, but are rare in Northern Ireland and Scotland.

How Do Tornadoes Form?

Tornadoes are nature's most violent storms. Most UK tornadoes are associated with cold fronts or non-supercell thunderstorms. The most destructive tornadoes spawn from supercell thunderstorms and develop in three stages.

STAGE ONE

Sunshine heats the Earth, which in turn heats the air near to ground level. Localised pockets of air become warmer than their surroundings and begin to rise. Where these thermals rise to sufficient height, shallow cumulus clouds develop. If the temperature in the surrounding atmosphere decreases rapidly with height because of an unstable atmosphere, the thermals may rise to much greater heights, resulting in the development of much deeper, stronger currents of ascending air, or updraughts, and deep cumulus and then cumulonimbus, or thunder, clouds.

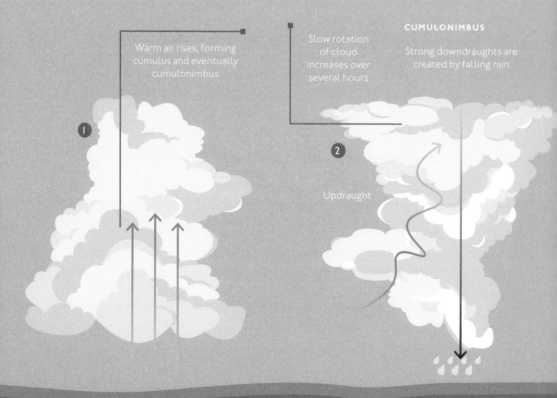

Warm air rises, forming cumulus and eventually cumulonimbus

Slow rotation of cloud increases over several hours

CUMULONIMBUS

Strong downdraughts are created by falling rain

Updraught

STAGE TWO

When the previous process occurs in an environment where winds increase strongly with height (strong vertical wind shear), the updraught in the thunderstorm may begin to rotate. This happens because the strong wind shear creates a horizontal spin in the atmosphere. Strong updraughts can then tilt this rolling motion into the vertical, so that the spin occurs about a vertical axis, a bit like the way a merry-go-round rotates around a central point. Thunderstorms that exhibit persistent and deep rotation are called supercells.

STAGE THREE

Downdraughts (descending currents of relatively cold, dense air) within a supercell storm help to concentrate the rotation and to bring it down to lower levels. Eventually the rotation may become so strongly focused that a narrow column of violently rotating air forms. If this violently rotating column of air reaches the ground, a tornado is born. The tornado is often made visible because of the presence of a condensation funnel – a funnel-shaped cloud that forms due to the much-reduced pressure within the tornado vortex. Dust and other debris made airborne by the intense winds can also help to make the tornado visible.

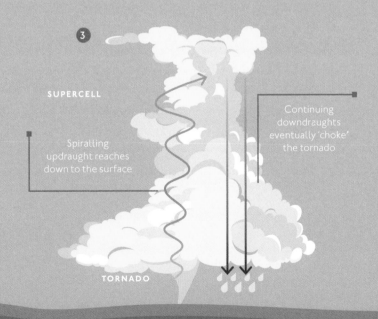

3

SUPERCELL

Spiralling updraught reaches down to the surface

Continuing downdraughts eventually 'choke' the tornado

TORNADO

How to Make a Tornado in a Jar

Fill a clear container, such as an empty jam jar, with water then add a few drops of washing-up liquid and food colouring. Tightly screw on the lid, swirl the container around in a circle lots of times and then stop. Inside you should see what a tornado looks like, which will slowly disappear as it reaches the top of the container.

BRITAIN'S STRONGEST-EVER TORNADO
209 to 240 miles per hour in December 1810 in Portsmouth

Britain's Most Notable Tornadoes

PORTSMOUTH TORNADO – 14 DECEMBER 1810

Believed to be Britain's strongest-ever tornado, giving a top wind speed of 209 to 240 miles per hour.

LITTLE LONDON TO COVENEY TORNADO – 21 MAY 1950

The tornado touched down at Little London in Buckinghamshire and travelled 107.1 kilometres to Coveney in Cambridgeshire, before covering a further 52.6 kilometres as a funnel cloud to Shipham in Norfolk and disappearing out to sea.

MULTIPLE LOCATIONS – 23 NOVEMBER 1981

The most tornadoes in one day, anywhere in Europe, with as many as 104 spawned by a cold front in the space of approximately five and a half hours. Thankfully, most were weak, and no deaths occurred.

SELSEY TORNADO – 8 JANUARY 1998

Most famous for demolishing one of astronomer Patrick Moore's observatories. The tornado lasted two to three minutes, covering a 3-kilometre track 600 metres wide. The insurance bill was £5–10 million.

BIRMINGHAM TORNADO – 28 JULY 2005

One of the strongest tornadoes in the UK in recent years. With a 7-kilometre track, it was estimated to have winds of 115 to 135 miles per hour.

Raining Fish

Despite the well-known saying, it has never actually rained cats and dogs in the UK, but it has, believe it or not, rained fish before. In 1666, a shower of fish was recorded in Cranstead, Kent:

> ♦A great tempest of thunder and rain, and,
> although no ponds about, two acres were scattered
> over with whitings of the size of a man's little finger.
> This occurred on the Wednesday before Easter ♦.

E. J. LOWE, IN *NATURAL PHENOMENA AND
CHRONOLOGY OF THE SEASONS*, 1870

And this was not a one-off event: in 1830, fresh herring fell on fields on Islay in the Hebrides, and in February 1859, dozens of fish fell in Mountain Ash, South Wales. The likely explanation is that waterspouts or tornadoes crossing bodies of water picked up the fish before depositing them again on the ground, much to the confusion of any onlookers.

Blood Rain

Science and technology have transformed our understanding of the weather, and we now look to meteorologists and other scientists to help explain natural phenomena. In days gone by, however, people did not always have easy access to such knowledge, and instead supernatural explanations were often put forward. One such phenomenon is blood rain, which is now believed to be caused by red dust or aerial spores of a green microalgae called *Trentepohlia annulata*. However, prior to the seventeenth century, it was believed that the deep-red rain was actually blood, and was a bad omen. Zeus was said to have twice caused blood rain to fall in Homer's *The Iliad*, and closer to home, the *Anglo-Saxon Chronicle* reported that 'there was a bloody rain in Britain. And milk and butter were turned to blood. And Lothere, King of Kent, died'. We should be wary, though, of thinking ourselves superior to our less-well-informed ancestors – there is still a lot that we do not know, despite all of our scientific achievements. And it is perhaps not so hard to imagine that people might believe that blood falling from the sky was not such a good thing.

What Is Space Weather?

Space weather describes changing environmental conditions in near-Earth space and is a consequence of the behaviour of the sun, the nature of Earth's magnetic field and atmosphere, and our location in the solar system. The active elements of space weather are particles, electromagnetic energy and magnetic fields, rather than the more everyday factors that make up our weather: water, temperature and air.

The sun may appear as a constant in the sky, but its energy varies greatly, largely driven by changes in its magnetic field. Magnetic fields, radiation, particles and matter ejected from the sun at speeds of up to 3 million miles per hour can interact with the Earth's magnetic field and upper atmosphere to produce a variety of effects – from beautiful aurora displays to disruption to satellites and telecommunications.

Most of the time the Earth's own magnetic field helps deflect the worst effects of solar activity. But not always; during periods of high solar activity, when there are more solar flares and coronal mass ejections, Earth experiences increased impacts. Extreme events that cause the most disruption can occur at any time during the eleven-year solar cycle.

The Met Office Space-Weather Department

The threat of space weather to national infrastructure, UK industry and the wider public is such that in 2011 it was added to the Government National Risk Register of Civil Emergencies. In response, the Met Office Space Weather Operations Centre was created to provide the critical information needed to help protect the country from the serious threats posed by severe space-weather events. They use numbered scales similar to those used to describe hurricanes or earthquakes to classify space-weather conditions and communicate the impact on people and systems, which can vary widely across the globe and depending on the technological system in question. For example, the UK power grid has much shorter and more highly connected transmission lines than those in North America, so it is less susceptible to space weather.

What Are Coronal Mass Ejections and Solar Flares?

A coronal mass ejection (CME) is the discharge of material from the sun into interplanetary space. If the material is directed towards the Earth, the event may result in a disturbance to the Earth's magnetic field and ionosphere, a part of the Earth's upper atmosphere that ranges across the mesosphere, thermosphere and exosphere. Solar flares, on the other hand, are sudden releases of energy across the entire electromagnetic spectrum near to the surface of the sun and visible to us on Earth as flashes of increased brightness. They are hard to predict, and the energy produced can be detected in Earth's atmosphere as soon as 8.5 minutes after they occur. CMEs often happen in the wake of solar flares. They can take days to reach Earth, carrying a local magnetic field from the sun, and their arrival time is the focus of space-weather forecasting.

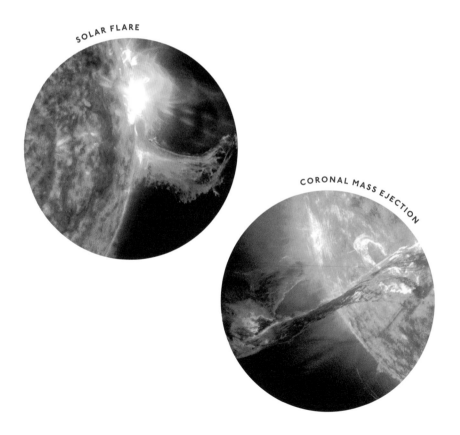

SOLAR FLARE

CORONAL MASS EJECTION

The Carrington Event

The largest solar event on record occurred between 1 and 3 September 1859. This was called the Carrington Event after Richard Carrington, the British astronomer who observed a huge solar flare the day before the sun subsequently discharged a CME that took just under 18 hours to reach the Earth. It was such a huge event that the aurora (*see* CME arrival *opposite*) was seen as far south as the Caribbean, and it caused widespread disruption to telegraph and telecommunications networks, as well as fires in some telegraph offices.

Nothing on this scale has happened in recent times, although we came relatively close in 2012 when a massive CME missed the Earth by approximately nine days – that is to say, the CME would have been directed towards the Earth if it had occurred nine days earlier in the sun's 25-day rotation around its axis. If we were hit by a CME comparable to the 1859 storm today, GPS, telecommunications and aviation would all be affected, causing massive disruption to our networked world in the twenty-first century.

The Northern Lights

The *aurora borealis*, also known as the Northern Lights, are one of nature's strangest and most impressive spectacles. The lights appear as large areas of colour, including pale green, pink, red, yellow, blue and violet, in the night sky in the direction due north. During a weak aurora, the colours are very faint and spread out whereas an intense aurora features greater numbers of brighter colours that can be seen higher in the sky. The lights generally extend from 80 kilometres to as high as 645 kilometres above the Earth's surface.

What causes the Northern Lights?

The Northern Lights occur as a consequence of solar activity and result from collisions of charged particles in the solar wind colliding with molecules in the Earth's upper atmosphere. Solar winds are charged particles that stream away from the sun at speeds of around 1 million miles per hour. When the magnetic polarity of the solar wind is opposite to the Earth's magnetic field, the two magnetic fields combine, allowing these energetic particles to flow into the Earth's magnetic North and South Poles. Auroras usually occur in a band called the annulus (a ring about 3,000 kilometres across) centred on the magnetic poles. The arrival of a CME from the sun can cause the annulus to expand, bringing the aurora to lower latitudes. It is under these circumstances that the lights can be seen in the UK.

Depending on which gas molecules are hit and where they are in the atmosphere, different amounts of energy are released as different wavelengths of light. Oxygen gives off green light when it is hit 96 kilometres above the Earth, while at 160 to 320 kilometres, rare, all-red auroras are produced. Nitrogen causes the sky to glow blue, and higher in the atmosphere, the glow can have a purple or violet hue.

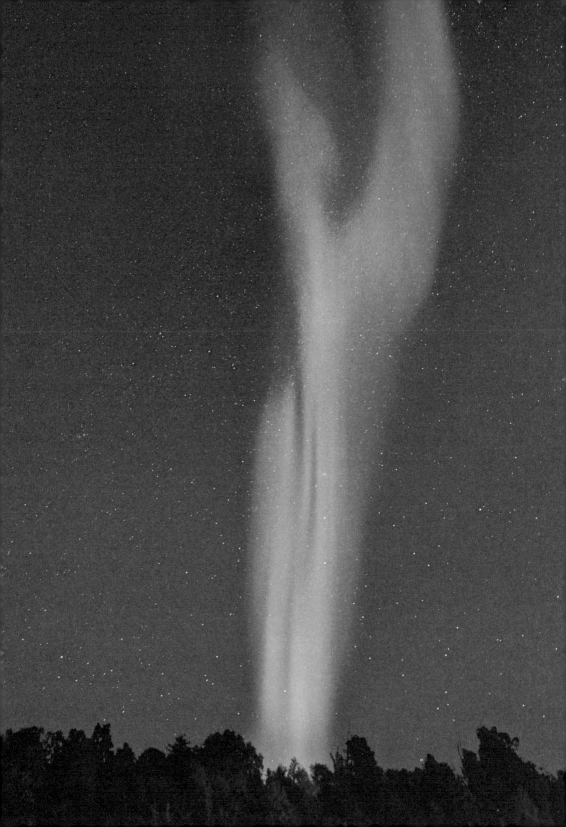

Which weather conditions are best to view the Northern Lights?

The best conditions to view the lights are when the sky is dark and clear of any clouds, as cloud cover ultimately blocks the view of the light. Ideally, the lights will be best viewed away from any light pollution, in remote areas, facing the northern horizon – north-facing coasts produce some of the best viewing locations.

Where can you see the Northern Lights in the UK?

The Northern Lights are best witnessed the further north you are but can be seen if the conditions are right in Scotland, the north of England and Northern Ireland. However, under severe space-weather conditions, the lights can occasionally be witnessed throughout the UK.

The Desert Comes to the UK

When strong winds blow over deserts such as the Sahara, dust and sand can be lifted high into the atmosphere and carried hundreds or even thousands of kilometres, sometimes reaching as far as the UK. If rain then falls, it will wash out the dust from the atmosphere, evaporate and leave a residue of fine sand or dust – for example, on surfaces such as car windscreens. This normally happens once or twice a year somewhere in the UK, although it is usually fairly localised, rather than being spread across large areas. However, an extreme example of a widespread and substantial dust and sand fall occurred on 1 July 1968.

Thunderstorms in Algeria at the end of June lifted huge quantities of Saharan dust into the atmosphere. It was then transported on a southerly wind in a layer between 3,050 and 5,200 metres from the ground. This layer of dust crossed Europe and remained relatively intact until it arrived over the UK, where it clashed with significantly cooler and moister Atlantic air. The boundary between the hot, dusty desert air and the cool, moist Atlantic air stretched from Devon in the south to Teesside in the north. Along that boundary, severe thunderstorms developed and continued for more than 24 hours in places, resulting in devastating flooding, frequent lightning, tennis-ball-sized hail (some of the most widespread and damaging hail in UK history), day darkness (*see below*) and four fatalities (three from lightning and one from flooding).

Not only did the dust lead to abnormally dark skies, it is also thought to have contributed to the extreme rain and hail associated with the thunderstorms due to cloud seeding. (Water vapour in the atmosphere doesn't tend to condense into raindrops by itself; it normally collects around naturally occurring salts, dust or aerosols. These are known as cloud seeds.) Most places south of a line from Lancashire to Yorkshire were affected by this dust fall – it's allegedly one of the most widespread in the UK for 200 years.

A Red Sky During the Day

We are familiar with red skies in the evening and morning, but on very rare occasions we can also experience them during the middle of the day, when Saharan dust in the atmosphere turns the skies vibrant red or orange. On 16 October 2017, precisely thirty years after Michael Fish's so-called 'hurricane' (*see* 'Stormy'), ex-hurricane Ophelia hit the UK. This was unusual for several reasons. Although ex-hurricanes do occasionally arrive in the UK, Ophelia had already taken an unusual track. Forming close to the Azores and strengthening to a category 3 hurricane by 14 October, Ophelia was the easternmost Atlantic hurricane of that strength on record. Ophelia then moved so quickly north-east towards the UK and Ireland that it technically remained a hurricane until just a few hours before it struck Ireland as their worst storm for 50 years. En route, Ophelia picked up a combination of Saharan dust and smoke from Portuguese wildfires to help turn skies an eerie orange colour across parts of southern England, providing us with an Earthbound glimpse of what life on Mars might be like.

Radioactive Rain

Although many of the rare and one-off weather events we experience are natural in origin, human activity can also lead to unique weather outcomes. The Chernobyl nuclear disaster was one such terrible example. The radioactive plume released from the reactor in the immediate aftermath of the accident was carried south-eastwards and then south and north over Europe and the UK. Heavy rain in April and May 1986 deposited large quantities of radioactive caesium and iodine into the upland peat soils of the north and west of the UK, which then accumulated in the grazing sheep.

Approximately 9,800 farms in Northern Ireland, Scotland, North Wales and Cumbria were subjected to livestock movement restrictions as a result. All the animals on these farms had to be monitored for levels of caesium-137 prior to being moved down from the fells for sale. Animals that registered traces that were too high were sent to other pastures, including Dartmoor, until such times as the levels of radiation in their systems had decreased and they were then able to be sold – this was usually just a matter of a few weeks. The peak of radiation in Welsh lambs occurred in 1992, and it was not until 2012, twenty-six years after the disaster, that radiation levels had dropped sufficiently to release the final 344 farms in North Wales and eight farms in Cumbria from restrictions.

ESTIMATED SPREAD OF RADIOACTIVE CONTAMINATION
FROM CHERNOBYL 26 APRIL–5 MAY 1986

Who Turned the Lights Out?

Day darkness is a very rare phenomenon that can occur when a number of meteorological factors coincide, including thick clouds, fog and smoke. The effect can be so dramatic that it seems like day has turned into night and artificial lights are required.

The example from 1 July 1968 (see page 236) was a combination of Saharan dust and severe thunderstorms. An even more famous day-darkness event occurred in London on 16 January 1955 when a belt of extreme darkness passed over the capital just after midday and the light levels fell to less than one-thousandth of what would be expected on a clear January day. Thousands of people contacted the emergency services, and newspaper switchboards were deluged with calls from scared and confused people, many of whom thought something terrible was under-way. In fact, the day darkness was another example of human activity interacting with the weather in an unusual fashion. London was still badly affected by smoke pollution at the time. In the morning, some of this smoke was carried by a south-easterly wind towards the Chilterns, where it was stopped by a layer of warm air. A strong north-westerly wind then set in, moving the smoke back towards the city in thick layers, some-times up to 1,000 metres deep and blocking out any light from the sun. Although pollution in our cities is still a problem today, the move away from the domestic burning of coal means such an extreme day-darkness event as this one is much less likely now.

FREAKY

Heat Bursts

On 25 July 2019, the UK's hottest day on record, a rare heat burst occurred. Just after 8pm, in less than an hour, the temperature temporarily rose by 10°C in parts of Lincolnshire. This was due to a downdraft of hot and dry air associated with a decaying thunderstorm.

Heat bursts are rare because atmospheric conditions have to be just right. Sometimes, when a thunderstorm dissipates and moves away, it leaves a layer of much colder air high in the sky in its wake. Cold air is more dense than warm air, so this layer starts to fall to the ground. As it descends it is subjected to friction and increased air pressure, which helps to compress and rapidly warm the air as well as dry it out. It also picks up momentum as it falls, propelling it until it hits the ground as a hot and dry wind.

A HEAT BURST BETWEEN 8 AND 10PM

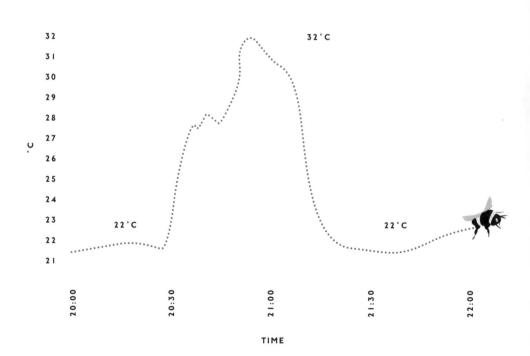

TIME

WATCH THE WEATHER WITH US

Continue the journey through our weird and wonderful elements

Join a global community of weather watchers – and record observations from wherever you are with **WOW**, the Weather Observations Website.
WOW.METOFFICE.GOV.UK

Explore the treasures of the **Met Office digital archives**, dating back to 1290!
DIGITAL.NMLA.METOFFICE.GOV.UK

Dive deep into our world-leading **weather and climate data**…
METOFFICE.GOV.UK/SERVICES/DATA

Share your own weather wonders using **#VeryBritishWeather** and tagging us with the handles below:

@METOFFICE

@METOFFICE

Acknowledgements

This book was the collective effort from many different experts at the Met Office who we'd like to thank here, starting with main contributors Aidan McGivern, Bonnie Diamond and Oli Claydon. Catherine Ross and Mark Beswick from the National Meteorological Library and Archive gave their time and a vast amount of historical archive material. Frank Barrow's scientific advice – and that of his colleagues Jodie Ramsdale and John Gunn – is invaluable, as is Dan Suri's comprehensive, expert feedback. Mark McCarthy and Mike Kendon at the Met Office National Climate Information Centre applied their excellent expertise, aided by Nicky Maxey and Grahame Madge.

Completing the book during both a busy winter for weather and a pandemic would not have been possible without a strong support network of family, and colleagues like Dave Britton, Jo McLellan, Sarah Fysh and Alex Deakin.

Thank you to Katie, Molly, and Daisy McGivern and the Diamond family for their love and support; to Monica and Mike McGivern for their inspiration and education; to Gilly Claydon for her unwavering support – she would have been very proud of her son's contribution to this book. And to Hugh McNulty, for his patience and adopted love of the weather.

With thanks to Emma Smith, Amandeep Singh, Catherine Ngwong and the rest of the team at Ebury – and Paul Murphy for writing.

Thank you to all colleagues from across the Met Office, including Michael Robbins, Ian Simpson, Graeme Anderson, Matthew Lehnert, Melyssa Wright, Richard Martin, Ian Watkins, Rachael Newton.

We are clearly better together.

 **Met Office**